EMPIRICAL-STATISTICAL DOWNSCALING

EMPIRICAL-STATISTICAL DOWNSCALING

Rasmus E Benestad, Inger Hanssen-Bauer
The Norwegian Meteorological Institute, Norway

Deliang Chen
University of Gothenburg, Sweden

NEW JERSEY • LONDON • SINGAPORE • BEIJING • SHANGHAI • HONG KONG • TAIPEI • CHENNAI

Published by

World Scientific Publishing Co. Pte. Ltd.
5 Toh Tuck Link, Singapore 596224
USA office: 27 Warren Street, Suite 401-402, Hackensack, NJ 07601
UK office: 57 Shelton Street, Covent Garden, London WC2H 9HE

British Library Cataloguing-in-Publication Data
A catalogue record for this book is available from the British Library.

EMPIRICAL-STATISTICAL DOWNSCALING

ISBN-13 978-981-281-912-3
ISBN-10 981-281-912-6

Typeset by Stallion Press
Email: enquiries@stallionpress.com

Printed in Singapore.

PREFACE

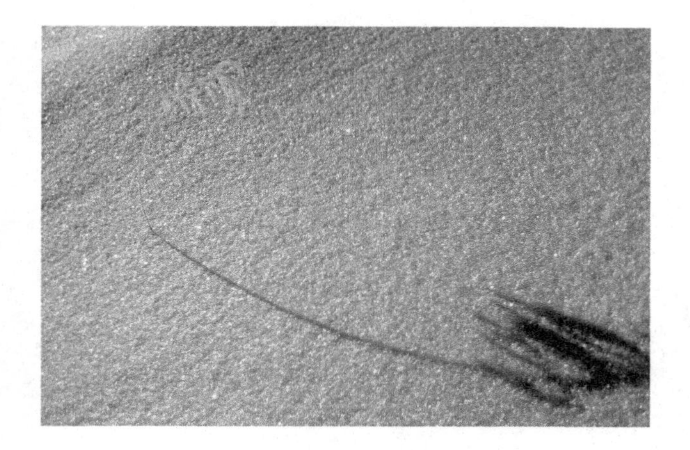

This compendium has been prepared for the benefit of the AMAP workshop in Oslo May 14–16, 2007, the Norklima-project (funded by the Norwegian Research council), the IPY-project EALÁT, Norwegian–Chinese collaborations, and a summer course organized by the Department of Meteorology and Climatology (EU FP6 Madame Curie action, STATME "Summer School in Statistical Downscaling"), University of Lodz, Poland, June 18–22, 2007. As climate change appears to have increasingly greater effect on local climate, the question of implications for the ecosystem and society becomes more and more pressing. It is especially true for "impact studies" which rely on local climate parameters as inputs for calculating the effect on the biosphere or on society. Another motivation behind this compendium is the notion of the need for an illustrated and easily read comprehensive text on empirical–statistical downscaling, starting from the basic principle and containing plenty of working examples. The compendium is based on numerous reports on downscaling and previous notes used in teaching statistics at the University of Bergen and Gothenburg. We hope that this compendium also will help researchers working on impact studies.

CONTENTS

Chapter 1

INTRODUCTION

1.1 History

The first account of empirical or statistical downscaling can be traced to Klein (1948), and a brief account of the early history of statistical predictions can be found in Klein and Bloom (1987). The first references to statistical downscaling activity is from numerical weather forecasting, and was then referred to as "specification."

In the early 1980s (Baker, 1982; Kim *et al.*, 1984) statistical downscaling was referred to as "Statistical Problem of Climate Inversion." However, similar techniques, so-called "Model Output Statistics" (MOS) and the "prefect prog" approach (Wilks, 1995) have been used in numerical weather forecasting since the early 1970s (Baker, 1982).

One reason why downscaling is a relatively young science is that it relies on the presence of global climate models, which themselves represent recent advances in the climate science community (Fig. 1.1).

There have since then also been some nice reviews of empirical–statistical downscaling (ESD) work (Fowler *et al.*, 2007; Christensen *et al.*, 2007; Wilby *et al.*, n.d.; Houghton *et al.*, 2001; von Storch *et al.*, 2000). However, these have been concise and not sufficiently comprehensive on the right level necessary for teaching purposes, since they tend to assume *a priori* knowledge which a student who tries to learn the subject may not have. Furthermore, these reviews have not captured some of the latest progress in this field, i.e. from works carried out in Scandinavia and Eastern Europe. Hanssen-Bauer *et al.* (2005), however, provided a review of the ESD-based work from Scandinavia. Another recent work (Graham *et al.*, 2008) summarized the ESD works for the Baltic Sea Region.

Fig. 1.1. The glasses here symbolize the techniques which provide a more detailed picture, an anology to downscaling.

Most of the literature on ESD has focused on Europe, although there are some publications with an North American focus (Lapp *et al.*, 2002; Easterling, 1999; Schoof and Pryor, 2001; Salathé, 2005), as well as for Australia/New Zealand (Kidson and Thompson, 1998), Africa (Reason *et al.*, 2006; Penlap *et al.*, 2004), and Southeast Asia (Oshima *et al.*, 2002; Das and Lohar, 2005). This compendium, however, will not dwell with the regional aspects, as the focus here will be on the methods rather than the results.

There is a number of reports and papers on ESD, but the various contributions are scattered across a large number of report series and journals. Here, we want to collect some of these in one volume, with a proper indexing and table of contents in order to make it easy for the readers to look up the topics and use the text as a compendium.

We also want to have a text which bridges the theoretical aspects with practical computer codes of problem solving. It is also our intention that the compendium provides some new and good illustrations for ESD and supplement the discussions with practical examples. Finally, we will attempt to divide the text into more readable main section and more detailed mathematical treatment in separate colored boxes.

1.2 The Lay-Out

Each chapter will contain an introduction of the concept, a theoretical discussion, illustrations, and a description of the mathematical notations. We will provide some practical examples, and a number of problems are given at the end in order for the reader to test her/him-self to see if the most important concepts are understood. The text will also include some mathematical treatment, but if it is above the most basic level, it will be separated from the main text and be placed in its own text box. The problems marked with an asterisk ("*") are advanced questions which require some analytical deduction.

There will be specific examples of how relevant concepts are formulated in the `R`-environment (Gentleman and Ihaka, 2000). The `R`-environment has a very nice searchable on-line help feature,[1] and is a freely available open-source numerical tool available from the Internet site `http://cran.r-project.org` (henceforth referred to as the "CRAN site" or just "CRAN"). Several `R`-packages have been compiled which can be used for ESD or general climate analysis, notably `clim.pact` (Benestad, 2004b, 2003a), `anm` (Imbert and Benestad, 2005), and `iid.test` (Benestad, 2004d, 2003d). These packages are available from CRAN under the link labeled "contributed." A fourth package which may be relevant to some studies is the package `cyclones` (Benestad and Chen, 2006). These packages will be used in the examples in the text below.

1.3 Concepts and Definitions

1.3.1 *The problem: What is downscaling and why downscaling*

1.3.1.1 *What is downscaling?*

The very basic question to be addressed first is *What is downscaling*? Here, we will define downscaling as *the process of making the link between the state of some variable representing a large space* (henceforth referred to as the "large scale") *and the state of some variable representing a much smaller space* (henceforth referred to as the "small scale").

Another view of ESD is that it basically is just an advanced statistical analysis of the model results.

[1]Type "`help.start()`" and a help manual appears in the browser.

The large-scale variable may for instance represent the circulation pattern over a large region whereas the small scale may be the local temperature as measured at one given point (station measurement). Figure 1.2 shows a map of correlation between the sea level pressure (SLP) and the North Atlantic Oscillation (NAO) index, bringing out the essence of the large-scale conditions: the SLP-field is correlated with SLP over Lisabon (Portugal) and anticorrelated with the SLP measured in Stykkisholmur (Iceland).

Another important characteristic, which we will return to later, is that the large-scale variable varies slowly and smoothly in space, which is reflected in the smooth correlation contours.

The small-scale variable may be a reading from a thermometer, barometer, or the measurement made with a rain gauge (Fig. 1.3). It is crucial that the link between the large-scale and the small-scale is *real* and physical, and not just due to a statistical fluctuation, coincidence, or an artifact of the statistical methods employed. ESD assumes an implicit and fundamental link between the two scales.

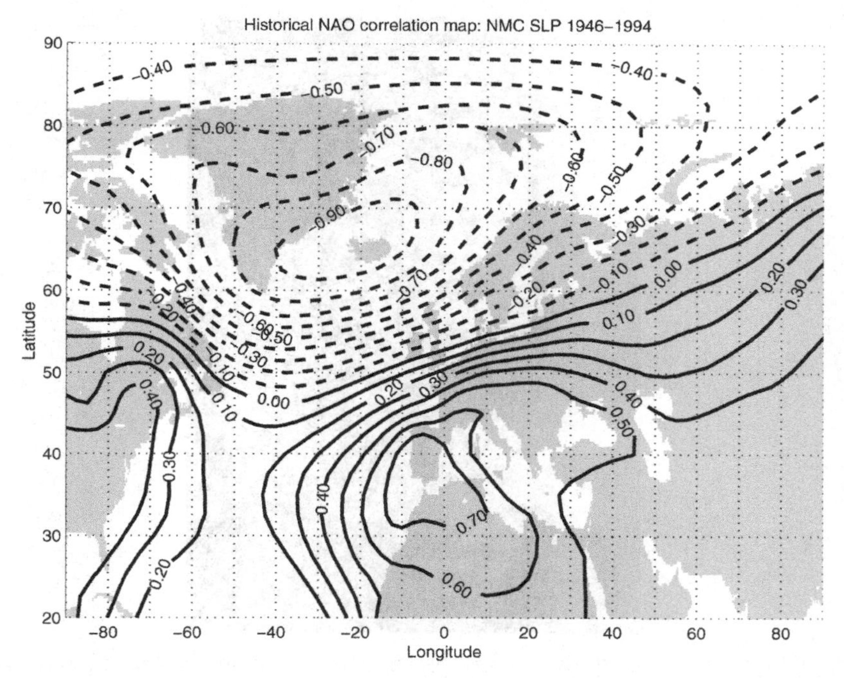

Fig. 1.2. Example of a large-scale variable, shown as a correlation map between the SLP and the NAO index.

It is important to distinguish the two concepts *large-scale* and *large volume/area*. The two are not necessarily the same, as a large volume may contain many noisy and incoherent small-scale processes.

One crucial question is whether the variable representing the large-scales variable used in ESD varies slowly and smoothly. If it does, then the spatial scale is large, but if there are pronounced changes over small distances, then the spatial scale is smaller. Thus, if the local variable is taken inside the space over which the large-scale is defined, then a significant part of the space necessarily covaries (i.e., has a high correlation) with the small-scale variable.

In a sense, the term "small-scale" may be a bit misleading, as we mean "local" rather than a process which only involves small spatial scales. In fact, the local process must be associated with large-spatial scales for downscaling to be possible.

To illustrate the concept of scales, let us consider a region which has frequent scattered cumulunimbus clouds (convective clouds). The rainfall

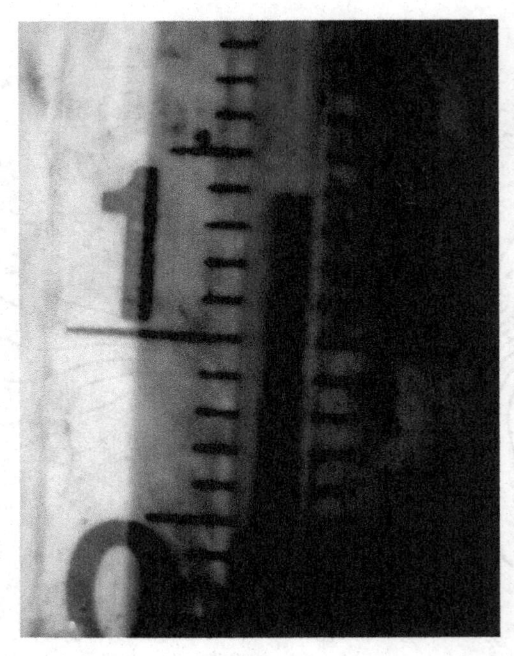

Fig. 1.3. Example of a small-scale variable: *in-situ* temperature readings representing the conditions at one particular location.

at one point is determined by the local presence of the cloud and its characteristics rather than the properties over a larger area. Thus, the spatial scale of the cloud process does not increase by looking at larger area or volume.

The essence of ESD is to identify synchronized or "matching" time behavior on large- and small scales, hence practical ESD focuses on the time dimension. We will henceforth refer to the character of temporal variation, which in mathematics can be described as a *function of time*, as "time structure." The implication of similar time structure on different spatial scales is a high temporal correlation.

Figure 1.4 illustrates the concept of time structure: the upper panel shows some made up time series whereas the lower panel shows real observations. The x-axis represents the time, running from left to right, and the data are shown in chronological order (time series). The time series with coherent variations in time have similar time structures, although their magnitude may be different.

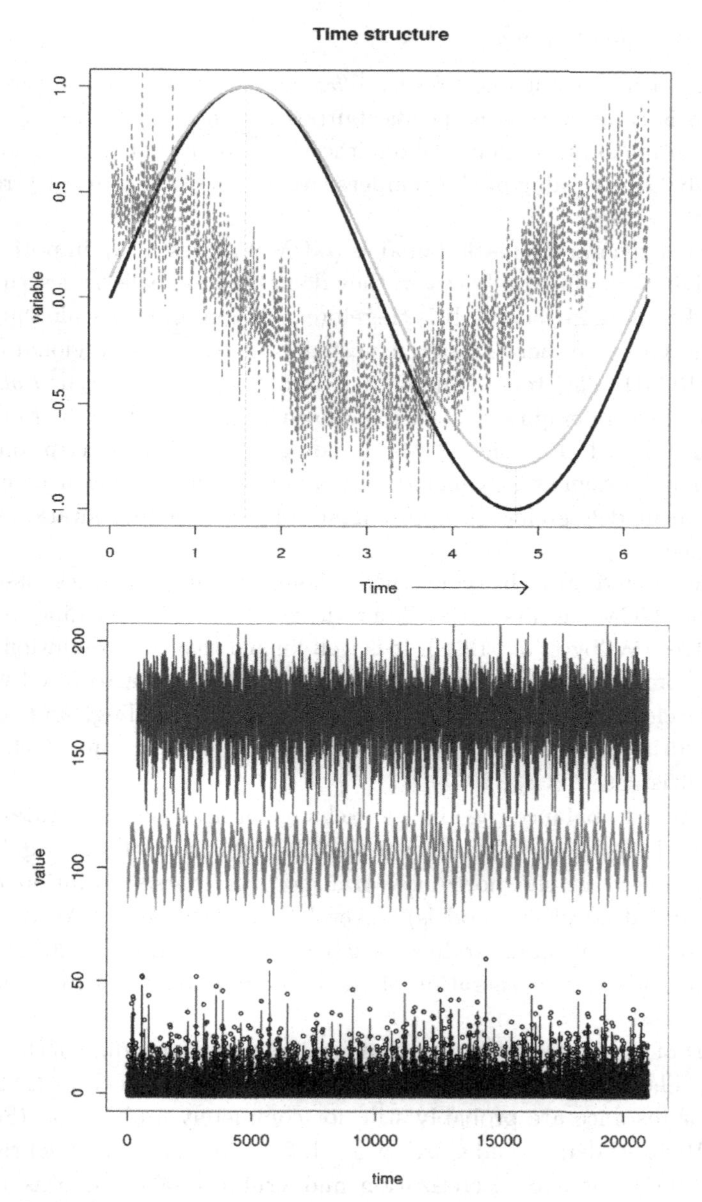

Fig. 1.4. Examples of "time structure." There the black and gray curves in the upper panel have similar time structure, but vary in amplitude. The red curve has a different time structure. Lower panel shows typical time structures from the real world.

1.3.1.2 *Why downscaling?*

The second important question is: *Why downscaling*? The answer to this question is connected to a specific purpose, such as using global climate models to make an inference about the local climate at a given location. The global mean value of the temperature is usually not directly relevant for practical use.

Global general circulation models (GCMs) represent an important tool for studying our climate, however, they do not give a realistic description of the local climate in general. It is therefore common to downscale the results from the GCMs either through a nested high-resolution regional climate model (RCM) (Christensen and Christensen, 2002; Christensen *et al.*, 2001, 1998) or through empirical/statistical downscaling (von Storch *et al.*, 1993; Rummukainen, 1997). The GCMs do not give a perfect description of the real climate system as they include "parameterizations" that involve simple statistical models giving an approximate or *ad-hoc* representation of sub-grid processes.

One should also be concerned about the uncertainties associated with the GCM results as well as those of the downscaling methods themselves (Wilby *et al.*, 1998). It is well known that low-resolution GCMs are far from perfect, and that they may have problems associated with for instance cloud representation, atmosphere–ocean coupling, and artificial climate drift (Bengtsson, 1996; Anderson and Carrington, 1994; Treut, 1994; Christensen *et al.*, 2007).

In order to balance the air–sea exchange of heat and freshwater fluxes, some GCMs also need to employ the so-called "flux correction" (e.g. because of a mismatch in the horizontal transport in coarse-resolution oceanic models and atmospheric models). Several state-of-the-art GCMs do not use flux correction but often produce *local* biases (Benestad *et al.*, 2002) despite giving a realistic representation of the climate system on continental and global scales.

Part of the problems are due to incomplete understanding of the climate system. The important mechanisms causing variability such as ENSO and NAO for instance are probably still not completely understood (Sarachik *et al.*, 1996; Anderson and Carrington, 1994; Philander, 1989; Christensen *et al.*, 2007). Due to discretization and gridding of data, it is unlikely that the global GCMs will simulate local details realistically (Crane and Hewitson, 1998; Zorita and von Storch, 1997; von Storch *et al.*, 1993; Robinson and Finkelstein, 1991).

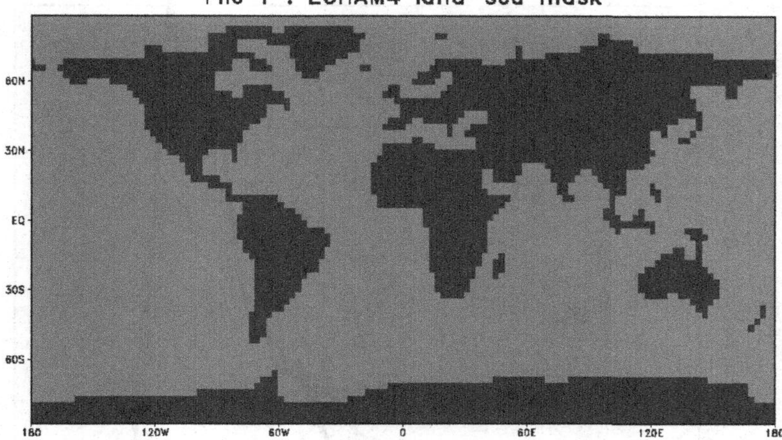

Fig. 1.5. An example of land–sea mask of a general circulation model (GCM) with ~2° × 2° spatial resolution (T42). Notice that Italy and Denmark are not represented in the model.

However, because a wide range of global GCMs predict observed regional features (e.g. the NAO, ENSO, the Hadley Cell, atmospheric jets), it is believed that the GCMs may be useful for predicting large-scale features.

Global climate models tend to have a coarse spatial resolution (Fig. 1.5), and are unable to represent aspects with spatial scales smaller than the grid box size. The global climate models are also unable to account for substantial variations in the climate statistics within a small region, such as the temperature differences within the Oslo region (Fig. 1.6). In principle, downscaling could be applied to more general settings, such as for instance relating the local wind speed to the SLP between two points.[2]

It is important to keep in mind the limitations of statistical downscaling, especially when applied to model results from Green House Gas (GHG) integrations using GCMs. The statistical models are based on historical data, and there is no guarantee that the past statistical relationships between different data fields will hold in the future. In the model stationarity section, we have applied a simple sensitivity test to see

[2]If all on the same line as the point where the measurement is made, then the SLP difference represents the large-scale geostrophic wind while the local wind measurement the local scale wind.

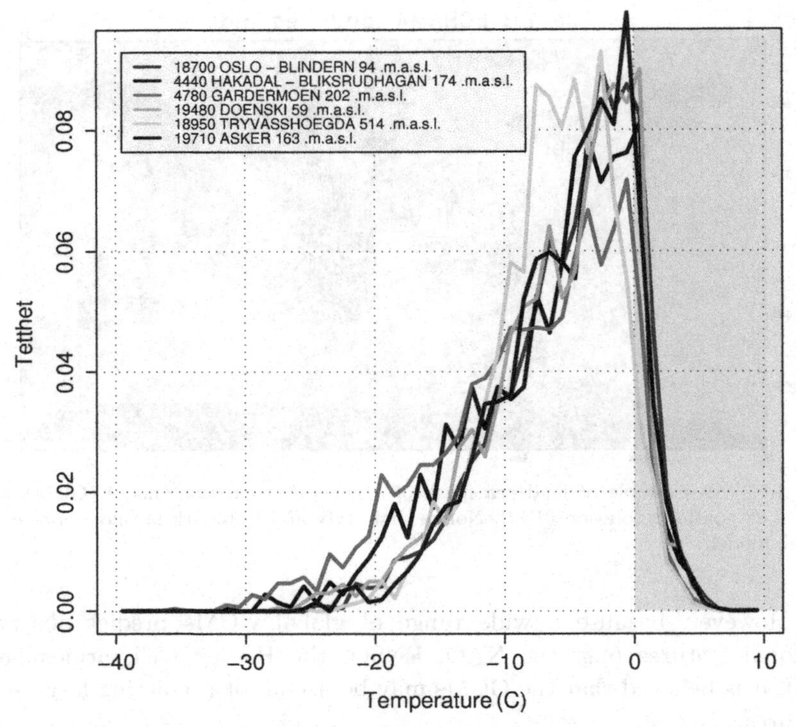

Fig. 1.6. Histograms of daily February–March minimum temperatures (TAN) for a number of locations around Oslo, Norway, all within a radius of 30 km from each other.

if the statistical model can be extrapolated to situations where the northern hemisphere generally is warmer than during the model training period.

The downscaling stage may introduce additional errors, despite a general added value, but systematic model biases may also severely degrade the downscaling performance. Table 1.1 gives a brief list of typical shortcomings associated with various models and analyses used in climate research.

1.3.2 *Notations*

In part of the scientific literature, the process of inferring information about the small-scale, given the large-scale conditions by the means of a statistical model, is referred to as "statistical downscaling" and in some publications as "empirical downscaling," with identical meaning.

Table 1.1. A list of typical drawbacks associated with the different models commonly used in climate research (from Imbert and Benestad (2005)).

GCMs	May have systematic biases/errors. Unable to give a realistic description of local climate in general. Processes described by parameterization may be nonstationary.
Nested RCMs	May have systematic biases/errors. Require large computer resources. Processes described by parameterization may be nonstationary. Often not sufficiently realistic description of local climate (Skaugen *et al.*, 2002b).
Empirical downscaling:	
Analog models	Cannot extrapolate values outside the range of the calibration set. Do not account for nonstationary relationships between the large-scale and local climate. Need a large training sample (often unsuited for monthly means). Do not ensure a consistency in the order of consecutive days.
Linear models	Assume normally distributed data. Tend to reduce the variance. Do not account for nonstationary relationships between the large-scale and local climate.

In both cases statistical methods are being employed, and these models are calibrated on historical measurements (observations), or empirical data. Thus in order to capture both these aspects, the term "empirical–statistical downscaling" would be appropriate. Hence, the terms "empirical downscaling," "statistical downscaling" and "empirical–statistical downscaling" can be regarded as synonyms, and we will use the latter expression in this text, but will henceforth use the abbreviation "ESD" to mean "empirical–statistical downscaling."

Henceforth, we will use "empirical data" as a synonym for "historical data" or "measurements." Different choice of these terms will only reflect on the readability of the text, but all will have the same meaning.

We will use the term "calibration interval" or "calibration period" when referring to the batch of data used for training the statistical model. This will generally refer to a time interval for the past when data are available for both predictand and predictor (see definitions on p. 16). On the basis of these data, different weights are applied to the different series in the predictor in order to obtain the best fit, also referred to as the "optimal fit."

Different criteria, such as maximizing the correlation or minimizing the root-mean-square-error (RMSE), are used in different optimalization methods discussed in the next chapters.

The term "scenario period" will in general refer to the future, and means a part of predictor data which have not been used in calibrating the statistical models. Thus, this term is used for out-of-batch, or *independent* data, whereas the calibration data are *dependent* data.

One important concept in geophysics is the "mode." Modes refer to pronounced recurring spatial patterns in nature, but also have mathematical property that one mode has a very different character to an other. Usually, the modes are said to be *orthogonal*, which means that they are uncorrelated, independent, or represent different dimensions.

One may draw an analogy from physics, normal modes, and resonance[3] in which certain harmonics are present and others not, just because the resonance means that only a whole number wave lengths may fit within the system. We will come back to modes when discussing EOFs in the next chapters.

1.3.3 *Definitions*

For practical purposes and for the benefit of readers with a different background than geosciences, it is useful to make a number of definitions:

Definition: series

A series is defined as a sequence of values, or record of measurements, arranged according to some order, thus implying several measurements. The measurements do not have to follow a regular structure.

A special case is the *time series* containing a number of different values ordered chronologically. We will also make reference to "station series," which implies a time series of some quantity (e.g. temperature) measured at a weather/climate station.[4] In the `clim.pact`-package, there is a "class"[5] of objects called "station" object, and a number of functions which know how to handle these. There are functions to retrieve some sample station series

[3] http://en.wikipedia.org/wiki/Normal_mode.
[4] A climate station is an instrument measuring a climate parameter fixed at one location.
[5] A label which tells R how to handle the data/object. The data may be stored as a structure with different types of data.

(`data(oslo.t2m)`, `plotStation(oslo.t2m)`). The mathematical notation for a series is often $[x_1, x_2, \ldots]$.

Definition: vector

A vector is a mathematical term used to represent a series, and is represented by the symbol $\vec{x}$. A vector holds several values. The individual values stored within a vector can be referred to by using an index, and x_i is the ith value of $\vec{x}$. The vector is characterized by the number of values it contains, or the vector length n. In the `R` environment, a vector is constructed using the command "`x <- c()`," and the ith value of $\vec{x}$ can be extracted by the call "`x[i]`." We will use both the notation $\vec{x}$ and x_i in the discussion of vector quantities and series.

In the following, we should be careful to distinguish the different meanings of "dimension." One common meaning is the number of different physical or mathematical dimensions a vector can represent. For instance, "`x <- c(x,y,z)`" can be used to represent a point in a three-dimensional room, and "`x <- c(x,y)`" on a two-dimensional surface. Thus, the physical dimensions are determined by the vector length.

In the `R` environment, however, the concept of dimension is often used to describe the structure of the vector or a data object. The command used to define the dimension is `dim(X)`. Whereas a vector has one dimension (e.g.,"`x[i]`"), it is possible to store data as tables with two or more dimensions (e.g. "`X[i,j]`," "`X[i,j,k]`," or "`X[i,j,k,l]`"). Thus, the second meaning of "dimension" tells us how many indices we need to extract any data point stored in the data object.

Definition: matrix

A matrix (plural: matrices) is a mathematical (linear algebra) construction similar to the vector, but differs by having two dimensions (a vector has only one dimension). Henceforth, matrices will be represented by capital letters, X, and individual values may be extracted by using two indices, e.g. i, j. The ith/jth value of X is X_{ij}.

It is possible in `R` to rearrange three-dimensional data objects describing two spatial dimensions (x and y) and one spatial dimension (t), so that the first index describes all the spatial dimensions and the other the time dimension. This means that a data object with different indices pointing to different locations on a surface (x and y) as well as different times t may be treated as a matrix with only two indices ($r(x, y)$ and t).

Definition: field

A field is used henceforth to refer a parameter which are measured simultaneously at different locations, and over a period of time. Often, the quantity is *gridded*, which means that the values are presented with a systematic and regular spatial structure, typically on a mesh.

A field can be thought of as a stack of maps, one for each observation time (i.e. every day or every month). It is difficult to visualize all the information embodied in a field in a static 2D graphics, but it may be possible through animations and multimedia devices. It is also possible to show one instant, such as the map for a given time, or to show maps of summary statistics. Figure 1.2 shows one example of how to present a field graphically, presenting the correlation between the SLP and the NAO index.

Each data point (or grid box) is characterized by its coordinates and by corresponding index. The way to represent a field mathematically is through a matrix X. The `clim.pact`-package provides a "field" class for objects representing fields, and a bundle of functions which can handle these. The `R`-call `data(DNMI.slp)` retrieves a SLP field (Benestad, 2000), and `map(meanField(DNMI.slp),sym=FALSE)` plots the mean field $\bar{X}$.

Definition: total value

The total value is defined as the value given in a conventional scale. The total value is related to the *absolute temperature*, which is the temperature scale in degrees Kelvin rather than Celsius, but we will in this context refer to any measurement provided in a conventional unit and with conventional reference points (e.g. 0°C) as "total value," and the purpose of this definition is to distinguish from other ways of presenting a measurement, such as the "anomaly" and "mean value."

Definition: mean

The mean (also referred to as the "first moment") is defined as the sum of a batch of values divided by the size of the batch (n). Mathematically, the mean can be estimated using the expression: $\bar{x} = \frac{1}{n} \sum_i^n x_i$.

In geosciences, the mean may be over several measurements in time (temporal mean) of over a region (spatial mean). We will henceforth use the notation $\bar{x}$ to represent the temporal (time) mean of any variable x, and $\langle x \rangle$ to represent the spatial mean. For variables with many dimensions, including both spatial and temporal, we can write the total mean as the spatial mean of the temporal mean $\langle \bar{x} \rangle$.

Often the mean is used to estimate "monthly values." For instance, the January mean for one particular year is the mean of all the measurements made in that month (measurements may be the daily mean temperature and the mean is taken from January 1st to January 31st). Note, for precipitation, the monthly value often is the accumulated value for the whole month ($x_{\text{tot}} = \sum_i^n x_i$, also referred to as the "total"), which is the sum of all the (24-h) precipitation measurements.

A different and a more mathematically rigorous/theoretical way to estimate the mean is through the location parameter for the probability density function (PDF). In the more theoretical approach, the location parameter is represented by μ rather than $\bar{x}$. If $f(x)$ is the PDF for x, then the mean can be expressed as $\mu = \int_{-\infty}^{\infty} x f(x) dx$.

Definition: climatology

The climatology is normally defined as the average value of a meteorological/climate variable for a given time of year. Thus, the climatology for the surface temperature in the high-to-middle latitude northern hemisphere varies from being maximum in summer and minimum in winter. The climatology is also often referred to as the "normal." Henceforth the climatology is for a variable x written as x_c.

The climatology differs from the mean value $\bar{x}$ as the latter does not describe the seasonal variation in x, and it can be estimated through two different strategies: (1) by taking the mean value over all days representing the same time of the year (e.g. the mean over all the January monthly mean values); (2) by applying a least-squares fit (regression) of a number of harmonics (sinus and cosine series) to the total values.

Definition: anomaly

The anomaly is defined as deviation from the climatology. The mathematical definition of the anomaly of x is $x' = x - x_c$. In `clim.pact` there are two functions which estimate the anomalies for either a station series of a gridded field, `anomaly.station()` and `anomaly.field()`.

Definition: standard deviation

The standard deviation is a measure of magnitude of the variations between different values stored in $\vec{x}$, and can be estimated according to $s_x =$

$\frac{1}{n-1}\sum_i^n (x_i - \bar{x})^2$. The standard deviation is often used for scaling, or standardizing, a series.

A different and a more mathematically rigorous/theoretical way to estimate the standard deviation is through the scale parameter for the PDF, and the scale parameter is represented by σ rather than s_x. If $f(x)$ is the PDF for x, then the standard deviation can be expressed as $\sigma = \int_{-\infty}^{\infty} x^2 f(x)dx$.

The standard deviation is also a measure of the difference between two series: $s_{x,y} = \frac{1}{n-1}\sum_i^n (x_i - y_i)^2$.

Definition: predictor

The predictor is the input data used in statistical models, typically a large-scale variable describing the circulation regime over a region. The predictor will be referred to by the mathematical symbol $\vec{x}$ if it represents a single series (univariate), or as X if it contains several parallel series (multivariate), e.g. a field. The predictor is also known as "independent variable,"[6] or simply as the "input variable," usually written as the terms on the right-hand side of the equation:

$$\text{predictand} = f(\text{predictors}). \tag{1.1}$$

Definition: predictand

The predictand is the output data, typically the small-scale variable representing the temperature or rainfall at a weather/climate station. We will henceforth refer to the predictands as the quantity predicted by the mathematical symbol $\vec{y}$. The predictand is also known as "dependent variable,"[7] "response variable," "responding variable," "regressand," or simply as the "output variable," usually the terms on the left-hand side of the equation

$$\vec{y} = f(\vec{x}). \tag{1.2}$$

In the equation above, $\vec{y}$ represents the predictand whereas $\vec{x}$ is the predictor. Figure 1.7 provides a schematic illustration of Eq. (1.2), showing the relationship between the predictand and the predictor.

[6] http://en.wikipedia.org/wiki/Independent_variable.
[7] http://en.wikipedia.org/wiki/dependent_variable.

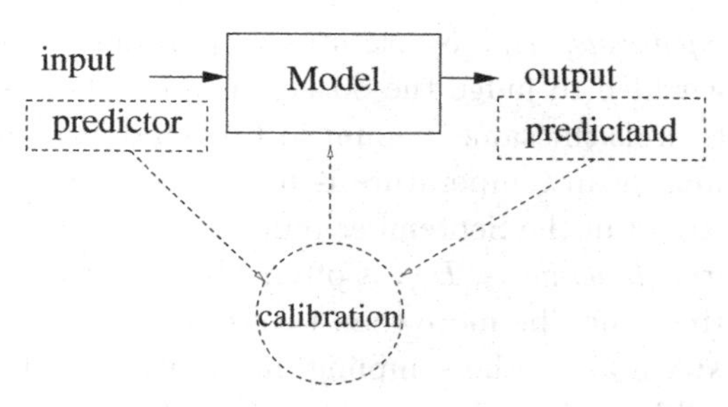

Fig. 1.7. Schematic illustration showing the relationship between predictor and predictand. Both are required for the calibration/training of the statistical model. The model in turn is able to calculate the predictand (output), given the predictor (input).

Predictors and Predictands

Linear downscaling assumes that local observations may be related to large-scale circulation patterns though a simple linear statistical relationship, such as $\vec{y} = M\vec{x}$ (Zorita and von Storch, 1997; von Storch *et al.*, 1993), $\vec{y} = MX$ (e.g. multiple regression) or $Y = MX$ (e.g. multivariate regression). In this case, the predictands and predictors contain several observations and are represented by vectors or matrices.

In the case of multivariate predictand, we will use the matrix Y to refer to the time series of $\vec{y}(t)$, where the vectors are given as the columns of Y. The two data fields Y and X contain data which are sampled at p and q locations, respectively, and over a time period with n measurements at each location:

$$\begin{aligned} Y &= [\vec{y}_1, \vec{y}_2, \dots \vec{y}_n], \\ X &= [\vec{x}_1, \vec{x}_2, \dots \vec{x}_n]. \end{aligned} \tag{1.3}$$

At time t, the data fields can be written as $\vec{y}_t = [y_{t1}, y_{t2}, \dots y_{tp}]^{\mathrm{T}}$. There are several techniques to find coupled patterns in climate data (Bretherton *et al.*, 1992), such as regression, CCA, and SVD methods.

ESD may involve one or more climate indices as predictors. When only one index variable is used as predictor, then this is referred to as "univariate" whereas when several parallel indices are used it is referred to as "multivariate" analysis.

Standard definitions in statistics

The *test statistics* is a parameter or quantity that in one way describes or represents the observations which we want to study.

The *null hypothesis*, H_0, constitutes a particular logical frame of reference against which to judge the observed test statistics. For example, a H_0 may be: "It rains the same amount in Bergen during September as in Oslo," "The global mean temperature is not influenced by the sunspots," or "There is no trend in the September temperature in Bergen."

The *alternative hypothesis*, H_A, is often "H_0 is not true," but in some complex cases, there may be more than two outcomes.

The *null distribution* is the sampling distribution of the test statistics given that the null hypothesis is true. This gives the range of values for the outcome of a test/analysis that is likely if H_0 is true. It is essential that we know the null hypothesis implied by the various methods. It is often a good idea to start the analysis with explicitly stating H_0, so that we know what we are looking for.

There are no definite answers in statistical analysis, but it is possible to make an estimate of the probability of data being consistent with a null hypothesis by comparing the data with the null distribution. Therefore, there is always a risk that an incorrect inference is made, as occasionally the improbable solution is the correct one. (For instance, it is improbable to find life in the universe, but despite the low probabilities, it is the correct explanation for planet Earth.)

We can refer to incorrect inference as errors, and thus classify them as following: Type I (probability: α) errors denote the false rejection of H_0, and II (probability: β) errors occur when H_0 is rejected when it is in reality true.

1.3.4 *Anomalies in ESD*

It is common to focus on anomalies in ESD work, and one common approach is to infer the anomalies, but combine these with the empirical climatology only in the final presentation of the results. It may be possible to by-pass systematic errors in the climate model (e.g. associated with representing the seasonal variations) by removing the climatology from the model results as well as the observations. In this context, there may be different types of "systematic errors."

One should take notice when climate models do not give an accurate representation of the annual cycle, as this is a sign that the model does not predict a correct response to well-understood cyclic variations in the external forcings (solar angle).

On the other hand, the local climatology is often determined by local physiography (local physical environment, or physical geography) unresolved by the climate models (Christensen *et al.*, 2007; Houghton *et al.*, 2001). An ESD based on anomalies combined with the empirical climatology will nevertheless produce a more realistic picture of the local climate. If the model results are poor, attention should be on improving the global climate model (or using a different one). Moreover, a completely wrong seasonal cycle may suggest that the model is not suitable for downscaling over the given region, whereas a constant bias (e.g. due to differences in elevation) is to be expected and will be corrected for through the ESD.

1.4 Further Reading

- Literature cited in this chapter.
- Linear algebra, matrices: Strang (1988), Wilks (1995), and Press *et al.* (1989).
- Houghton *et al.* (2001)[8] and Christensen *et al.* (2007).[9]

1.5 Examples

- One example of a *series* is $\vec{x} = [1, 2, 3, 4]$. In R, this is written as `x=c(1,2,3,4)`.
- *The total value* for the air pressure in Fig. 1.8 is 764 mm/Hg (this unit because the barometer is old, but a scaling factor of 1.333224 yields 1018.6 hPa).[10] Likewise the total value for the temperature reading in Fig. 1.3 is 14°C, which is the same as $273 + 14 = 287$ K (absolute temperature, not to be confused with the *total value*).
- *The mean* of the sequence $\vec{x} = [1, 2, 3, 4]$ with the length $n = 4$ is $\bar{x} = \frac{1}{n}\sum_i^n x_i = (1+2+3+4)/4 = 10/4 = 2.5$. In Fig. 1.9, the mean of daily mean air temperature (TAM) is shown as a blue horizontal line.
- *The climatology* of the TAM for Oslo–Blindern is shown as a red curve in the upper panel of Fig. 1.9.
- *The anomaly* is shown as the black symbols in the lower panel of Fig. 1.9, and the deviations from the mean is shown in blue. Below

[8] http://grida.no/climate/ipcc_tar/index.htm.
[9] http://ipcc-wg1.ucar.edu/wg1/wg1-report.html.
[10] http://www.paroscientific.com/convtable.htm.

Fig. 1.8. A barometer showing the air pressure reading near sea level.

is an example in R for finding the anomalies of a station object (here, one of the sample station series included in `clim.pact`):

```
> library(clim.pact)
> data(oslo.t2m)
> a <- anomaly.station(oslo.t2m)
> print(a)
> plotStation(oslo.t2m)
```

whenever we list lines of a computer script, we show the prompt ">" before the actual command. The user should not write the prompt, unless stated otherwise — the R-environment generates these.

1.6 Exercises

1. Explain the difference between the predictand and predictor.
2. Discuss the most important reason for carrying out ESD.
3. What is meant by "climatology"?
4. Discuss the general aspects of two different ways of estimating the climatology for a station series.
5. Explain what is meant by a "null hypothesis." Use `data(oslo.t2m)` and `data(bergen.t2m)`; how would you determine if these really

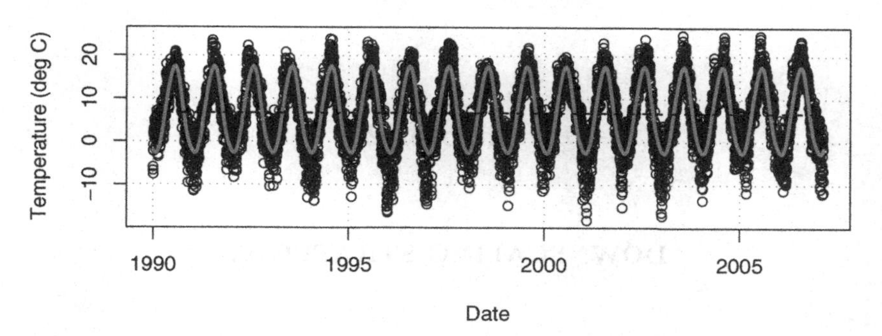

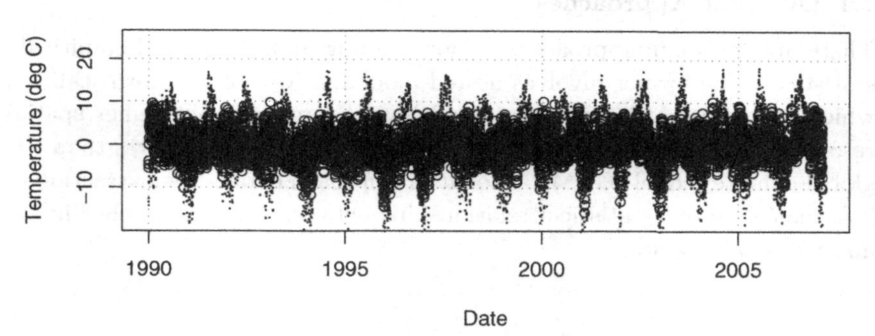

Fig. 1.9. Upper panel: Total value (daily mean temperature, TAM) in black, the climatology in red, and mean value in blue. Lower panel: the temperature anomaly in black (the difference between the black and the red curve in the upper panel), and the difference deviation from the mean value.

are different? Use the call `t.test()` to compare the mean January temperatures.

6. How would you model a steady increase in the temperature (simple trend analysis)?
7. Retrieve station series for Oslo, Bergen, Helsinki, Stockholm, and Copenhagen and use a simple linear regression (`lm`) to estimate the linear trends. Plot the results.
8. Plot the anomalies of the temperature in Oslo. Then plot the climatology.

* Discuss why there may be problematic to downscale a large-scale field which does not vary smoothly spatially.

Chapter 2

DOWNSCALING STRATEGIES

2.1 Different Approaches

There are two main approaches to downscaling: dynamical and empirical–statistical. The former involves nested modeling (dynamical downscaling), which makes use of limited area models with progressively higher spatial resolution that can account for more of the geographical features than the global climate model (GCM). The latter approach entails the extraction of information about statistical relationships between the large-scale climate and the local climate.

2.1.1 *Dynamical downscaling*

Dynamical downscaling is also referred to as "numerical downscaling" or "nested modeling." The dynamical downscaling approach provides an alternative to the statistical downscaling, but without assuming that historical relationships between large-scale circulation and local climate remain constant (Fig. 2.1).

In theory, these nested dynamical models are physically consistent representation of a small region of the atmosphere, and it is indeed remarkable that the dynamical climate models reproduce the main features of the climate as realistically as they do, considering that they are based on merely fundamental physical laws. One demonstration of regional climate models' (RCMs') merit is that they have been applied skillfully to different regions around the world. However, the dynamical models are not perfect and there are some drawbacks associated with

Fig. 2.1. There are examples in nature where the properties or character does not vary smoothly in space. One example includes minerals, crystals, etc.

dynamical downscale models, such as:

(i) The dynamical downscaling models are tuned for the present climate. Cloud schemes are parameterized and based on empirical relationships (Bengtsson, 1996; Heyen *et al.*, 1996), and the parameterization of cloud radiation is notoriously difficult to implement in climate and weather forecast models (Palmer, 1996). Furthermore, we do not know if these parameterization schemes will be valid in a global warming scenario even if they were appropriate for the present climate (same problem as stationarity in empirical–statistical downscaling (ESD), discussed below). This issue also concerns the GCMs, whose results are used as predictors for all the downscaling models.

(ii) The dynamical downscaling models are to date extremely expensive to run, and only a few integrations can be afforded. This inhibits the use of dynamical models for long integrations and extensive hypothesis testing.

(iii) Upscaling instabilities (Palmer, 1996; Lorenz, 1963, 1967) are filtered out: these may not be important in operational weather forecast integrations, which are integrated over a shorter period, but upscaling instabilities can cause inconsistencies in longer integrations between the model boundary values and the internal dynamics. The sea surface temperatures (SSTs) taken from the coupled-GCM run may also not be appropriate as boundary values in a nested model if these are sensitive to local ocean forcing (like Ekman pumping).

(iv) Schemes to counter numerical instabilities are usually needed due to the fact that the model consists of discrete values on a discrete grid. Furthermore, there are no perfect numerical intergation schemes, and results from dynamical models are subject to round off errors as well as numerical diffusion and inadequate conservation properties (Palmer, 1996; Press *et al.*, 1989).

(v) Concerns have also been raised about the effects of lateral boundary conditions resulting in ill-posed solution (Kao *et al.*, 1998; Mahadevan and Archer, 1998). Furthermore, a RCM may provide an inconsistent picture if the lower boundaries (sea surface conditions) are prescribed from a GCM experiment and coupled air-sea processes are present.

RCMs tend to inherit systematic errors from the driving models (Machenhauer *et al.*, 1998), as dynamic downscaling models may exaggerate the cyclone activity over the North Atlantic and therefore give excessive precipitation and warm biases in the northern Europe. Often the results from RCMs are not representative for the local climate, and statistical–empirical schemes must be employed to adjust the data in order to obtain a realistic description of the local temperature or rainfall (Skaugen *et al.*, 2002b; Engen-Skaugen, 2004).

In most cases, however, one would expect that the shortcomings of the dynamical and statistical models to produce different errors, and therefore a combination of the two methods may be particularly useful.

Dynamical downscaling will not be the focus of this text, but will only be referred to as a means to put ESD into perspective.

2.2 Philosophy Behind ESD

In the previous chapter, the essence of downscaling was explained as utilization of the link between different scales to say something about the smaller-scale conditions given a large-scale. The reason why downscaling is useful was also discussed. Here, we will elaborate on the merits of ESD as well as discuss some fundamental issues associated with modeling.

ESD can be used to provide the so-called *assimilation* of the predictions, which means that the results should have same statistical distribution (or probability density function, PDF) as the real data. In mathematics, this is also referred to as mapping the results of the empirical data so that they describe the same *data space* (a mathematical/statistical concept), not to be confused with producing geographical maps (this will be discussed later). It is important to keep in mind that the statistical distribution (PDF) for the predictor will determine the distribution of the downscaled results in ESD.

A common problem is deriving a representation for the local precipitation amount which is both non-Gaussian and often contains a large fractions of nonevents (no precipitation). The predictor, for instance sea level pressure (SLP), may on the other hand be characterized by a Gaussian distribution.

One solution to this problem may be to transform the predictand so that it becomes (approximately) linear (Benestad and Melsom, 2002). It is also possible to circumvent this issue by utilizing nonlinear techniques or applying weather generators. These techniques will be discussed in later chapters.

In climate change studies, one important question is what implications a global warming has for the local climate. The local climate can be regarded as the result of a combination of the local geography (physiography) and the large-scale climate (circulation). Until now, GCMs have not been able to answer this question since their spatial resolution is too coarse to give a realistic description of the local climate in most locations. Furthermore, local effects from valleys, mountains, lakes, etc. are not sufficiently taken into account to give a representative description (Fig. 2.2 provides an example of local effects, where fog forms over a lake on the lower plateau). It may nevertheless be possible to derive information for a local climate through the means of downscaling.

The local climate is a function of the large-scale situation X, local effects l, and global characteristics G, described mathematically as

$$y = f(X, l, G). \tag{2.1}$$

This formula will be a central framework for ESD and is inspired by von Storch *et al.* (2000). Here, X is the regional effects that not directly influenced by G. Although it is expected that variations in the global mean will imply changes in the regional climate, it is not necessarily guaranteed that a change in X will follow that of G systematically.

Fig. 2.2. Panoramic morning view of the Rondane mountain range.

2.2.1 *Considerations*

It is important that there is a strong underlying physical mechanism that links the large- and small-scale climate, as the lack of a physical basis cannot preclude the possibility of weak or coincidental correlations. Linear-empirical downscaling provides approximate description of the relationships between the spatial scales.

Furthermore, empirical downscaling facilitates a "correction" to the simulations, so that the observations and the downscaled simulations can be regarded as directly comparable.

A nice feature of empirical downscaling is that this method is fast and computationally "cheap." The drawback of empirical downscaling is that the locations and elements are limited to those of historical observations.

Advantages

(i) ESD is cheap to run, which means that we can apply these statistical models to results from a number of different coupled GCMs. We can therefore get an idea of the uncertainties associated with the GCMs.

(ii) ESD can be tailored for specific use, and the statistical models can be optimized for the prediction of certain parameters at specified locations, as for instance specified by customers. This makes the statistical model approach ideal for end users.
(iii) Dynamical downscaling models still have a low spatial resolution for some impact studies, and one may still have to apply some kind of downscaling/MOS technique to the dynamical model results.
(iv) The statistical models can be used to find coupled patterns between two different climatic parameters, and hence provide a basis for analyzing both historical data as well as the results from dynamical downscaling. An alternative approach to improve our physical understanding is to run model experiments with GCMs or nested models, however, this kind of numerical experiments requires substantial computer resources and is expensive.

Assumptions

Regarding the predictors, Hellström *et al.* (2001) argued that: "(1) they should be skillful in representing large-scale variability that is simulated by the GCMs; (2) they should be statistically significant contributors to the variability in predictand, or they should represent important physical processes in the context of the enhanced greenhouse effect; and (3) they should not be strongly correlated to each other."

The latter point here can sometimes be relaxed if methods used do not rely on the predictors being uncorrelated, and principal component analysis (PCA; see the discussion on EOFs in the next chapters) can remould the data so that the input to the ESD is orthogonal.

Nevertheless, if two input variables are correlated with the predictand during the calibration period, and only one responds to a climate change, then it is likely that the ESD will fail to provide a good indicator about a climate change. It is therefore important to have a good physical understanding about which predictors have a physical connection to the local variable, and it is important to limit the set of predictors to only those which are relevant.

This idea is discussed in von Storch *et al.* (2000), who list a number of criteria which must be fulfilled for ESD: "(1) The predictors are variables of relevance and are realistically modeled by the GCM; (2) The transfer function is valid also under altered climatic conditions. This is an assumption that in principle cannot be proven in advance. The observational record should cover a wide range of variations in the past; ideally, all expected future realizations of the predictors should be contained in the observational record; (3) The predictors employed fully represent the climate change signal."

Here, we will summarize these criteria into four necessary conditions which *must* be fulfilled in ESD:

(a) Strong relationship
(b) Model representation
(c) Description of change
(d) Stationarity

If any of these conditions are not fulfilled, then the ESD may be flawed and pointless. We will discuss each of these in more detail below.

Strong relationship

The basis of ESD is the assumption that there is a close link between the large-scale predictor and the small-scale predictand, thus a strong relationship. It is only when they to a large degree covary and have similar time structure that it is possible to use a predictor to calculate the predictand.

Model representation

ESD takes the predictor as given, and it is therefore important that the predictor is simulated well by the models. In other words, if the parameter taken as the predictor is unrealistic, then the ESD results will be wrong too.

Parameters such as geopotential heights, SLP, T(2m), and geopotential heights tend to be realistically captured by the GCMs (Benestad *et al.*, 1999; Benestad, 2001a), but the SST, which partly depends on the ocean dynamics, is not well-represented as the spatial resolution of the ocean models tends to the too coarse to describe the ocean currents which are important influences on the SST.

GCMs may also have shortcomings with respect to the description of the vertical profiles through the boundary layers or representation of humidity.

The question of the degree to which the predictor is representable also depends on time scale. Sometimes the monthly mean gridded values may give a reasonable description, whereas daily values may be more problematic.

It is not only the model results which is the issue here, but the gridded observations used for calibrating the models too. Large uncertainties in the gridded observations introduce difficulties in terms of model evaluations as well as in the matching of simulated traits to observed ones, thus leading to a weak relationship. Furthermore, errors in the gridded observations hamper skillful calibration of the statistical models.

Description of change

It is important that the predictor parameter responds to given perturbations in a similar fashion as the predictand, or the ESD results will not capture the changes. This can also be seen from the simple mathematical expression describing an ideal situation: $\vec{y} = F(X)$. If this equation truly is representative, the equality implies that y and $F(X)$ respond the same way. For a linear model, the function can be approximated by $F(X) \approx bX$, and the equation can be written as $y = bX$. Now, the function $F(.) \rightarrow b$ is taken as being stationary (does not change over time or value of X), meaning that y must change proportionally with X.

One example of ESD is the use of a circulation index such as the SLP to model the local change in T(2m). According to the first law of thermodynamics, any change in temperature dT/dt can be tied up to an energy input or loss (Q) and a change in the work applied: $cdT/dt = Q + pdV/dt$ (Zemansky and Dittman, 1981). This expression assumes a Langrangian reference frame, which can be expanded to the following equation of a fixed point (Eulerian reference).

Hence from the first law of thermodynamics, it is evident that the changes in temperature (dT/dt) is only partially described in terms of the pressure (p). The first law of thermodynamics can be expressed as

$$\frac{\partial T}{\partial t} + \frac{\partial T}{\partial x}\frac{\partial x}{\partial t} + \frac{\partial T}{\partial y}\frac{\partial y}{\partial t} = \frac{Q + pdV/dt}{c}. \tag{2.2}$$

Here, $\frac{\partial x}{\partial t}$ is physically the same as the air velocity along the x-direction and $\frac{\partial y}{\partial t}$ represent the motion along the y-direction. A mathematical shorthand for $\frac{\partial T}{\partial x}\frac{\partial x}{\partial t} + \frac{\partial T}{\partial y}\frac{\partial y}{\partial t}$ is $\vec{v} \cdot \nabla T$. Here, the velocity $\vec{v}$ should not be confused with the volume V, although the change in volume is related to the velocity field as the change of volume is the divergence of the flow: $dV/dt = \nabla \cdot \vec{v}$. After some rearrangement, Eq. (2.2) can be rewritten as

$$\frac{\partial T}{\partial t} = -\vec{v} \cdot \nabla T + \frac{Q + p\nabla \cdot \vec{v}}{c}. \tag{2.3}$$

We can take this even further by assuming that the air flow is in geostrophic balance (Gill, 1982). We use the notation $\vec{v}_g$ to indicate that we are referring to the geostrophic flow.

$$u_g = -\frac{1}{f\rho}\frac{\partial p}{\partial y}; \quad v_g = \frac{1}{f\rho}\frac{\partial p}{\partial y}. \tag{2.4}$$

The geostrophic wind can be expressed using a mathematical shorthand called the *curl* $(\nabla \times p)$ $\vec{v}_g = -\frac{1}{f\rho}\nabla \times p$, and our original expression can be written entirely in terms of the pressure, spatial temperature gradient, and heat flux Q:

$$\frac{\partial T}{\partial t}_g = \left(\frac{1}{f\rho}\nabla \times p\right) \cdot \nabla T + \frac{Q - p\nabla \cdot (\frac{1}{f\rho}\nabla \times p)}{c}. \tag{2.5}$$

Equation (2.5) shows why the SLP (p) is not a good single predictor choice for the temperature, as this excludes the effect of Q (e.g. radiative imbalance due to increased downwelling long-wave radiation, increased evaporation, or transport of latent heat through increased moisture) as well as the temperature advection due to changes in the large-scale temperature structure (spatial temperature gradient).

Another common case is the use of SLP to represent the ciculation regime and then to downscale precipitation P. In this case, we can use continuity equation for the atmospheric moisture ρ_w as a guide line, assuming that the evaporation is a function of temperature $E(T)$ describing a "source" term and the precipitation is a "sink" term: $\partial \rho_w/\partial t + \vec{v} \cdot \nabla \rho_w = E(T) - P$ (see the box above for the mathematical notations). The air flow $\vec{v}$ can be related to the pressure, as done in the box, but now one important component to the flow is the vertical ascent associated with convection, cyclones, or frontal systems.

Nevertheless, the important message is that the theoretical considerations suggest that a predictor including only the pressure can merely describe part of the precipitation, since it implicitly assumes that ρ_w is constant ($\partial \rho_w/\partial t = 0$).

Since the GCMs tend to provide a good description of the SLP, but SLP does not contain the "global warming signal," one may consider using it in combination with the global mean temperature to describe the variations in local conditions as well as the influence from a global warming (Hanssen-Bauer and Førland, 2000; van Oldenborgh, 2006).

Huth (2004) found that when using only SLP or 1000-hPa heights as the only predictor, ESD tends to lead to unrealistically low temperature change estimate.

Stationarity

The fourth important aspect to ESD is the issue of stationarity (Wilby, 1997). By this we mean that the statistical relationship between the predictor and the predictand does not change over time. In Eq. (2.6), the requirement is that l, which describes the effect of the local landscape/ geopgraphy, is constant. In other words, stationarity implies that the local landscape does not change.

Examples of landscape changes that may render the relationship between large- and small-scales nonstationary include deforestation, ascending tree line, encroaching urbanization, plowing up new fields for agriculture and introduction of irrigation, construction of dams to make nearby reservoirs, or a weather/climate station relocation.

Vegetation may change or snow may melt as a result of climate change, and may hence indirectly affect the local climate. For instance, global warming may result in a higher tree line.

To some degree, the complex coupling between the local climate and the vegetation may be captured by the empirical models if the future scenario follows a pattern seen in the past, but this cannot be guaranteed. Thus, there may be several reasons why l is not a constant.

The other aspect of nonstationarity in the expression $y = f(X, l, G)$ is whether the function G varies for some other reason. Here, we use l to represent local effects, and changes to G would entail a more global scale, but the reason for this kind of change is more unclear.

One interpretation of G is the effects on the local climate from a global change, not captured by GCMs due to shortcomings or biases, whereas X represents the predicted regional response.

Some examples may be that a climate change due to changes in the ocean circulation (e.g. the global mean temperature, the "thermohaline circulation" or the "global conveyor belt"), teleconnections, nonlinearities, changes caused by changes in the planetary-scale poleward heat, vorticity and mass transport (e.g. changes to the Hadley Cell and the polar ice-cover), and other forcings such as solar and volcanic eruptions.

If the contributions from l and G are constant, then we can rewrite Eq. (2.6) as

$$y = f'(X), \tag{2.6}$$

where $f'(\cdot)$ is a function that represents the effects of l and G in $f(\cdot)$. This is the equation discussed by von Storch *et al.* (2000) and this is the form that will be used henceforth in this text. Thus, ESD discussed henceforth only considers the relationship between the predictand y and the regional predictor X. The objective of the remainder of the compendium is to discuss various techniques to derive $f'(\cdot)$.

2.2.2 *A physics-inspired view*

Part of the discussion in Sec. 2.2 was inspired by physics and the fact that the left-hand side of an equation is by definition equal to the right-hand side. The examples above were based on the first law of thermodynamics, the continuity equation, or a geostrophic balance. For such cases, a linear model can in principle be employed to represent the relationship between all the terms on the left-hand side on the one hand, and all the terms on the right-hand side on the other, given a well-defined physical condition. These should have the same physical units if the equation is to represent a physical law (Fig. 2.3).

However, when relating small-scale phenomena to large-scale conditions, there may not be a strict one-to-one relationship. The small-scale may follow the large-scale conditions to some degree, but also exhibit a behavior independent of the large-scale situation. The independent behavior will not be captured by the predictor, and will henceforth be referred to as *noise*.

The part of the local variability related to the large-scale, may have different amplitude for different locations, and thus each location may have a systematic relationship with the large-scale. It is this systematic pattern that is utilized in ESD.

In several studies, predictors have been chosen so as to capture the most important physical aspects related to the local climate variable. Chen *et al.* (2005) and Hellström *et al.* (2001), for instance, used the

Fig. 2.3. The notion that all processes ultimately are governed by physical laws should inspire the thinking about which conditions that are important for the prediction of the resulting phenomena.

two geostrophical wind components, total vorticity, and the large-scale humidity at 850 hPa height as predictors. The former two can be associated with advection processes, the total vorticity can give an indication of the ascent of the air, and the latter describes the amount of water available for precipitation.

One way to implement the physics-inspired ESD can be to use large-scale gridded data to predict the local variable of the same parameter.

For instance, using the large-scale gridded T(2m) analysis to predict the local 2-m temperature measured at a given location more or less fulfils the criteria of nonstationarity and containing the essential signal. Benestad *et al.* (2007) argued that the large-scale precipitation from gridded reanalysis products and GCMs serve as a reasonably skillful predictor for the local rainfall. One drawback may be a weak relationship between precipitation on local and the large scales, but this shortcoming tends to be common for all predictors when the predictand is precipitation (Benestad *et al.*, 2007).

There is a sound physical reason behind the wintertime SLP pattern and precipitation for most locations in Norway, as it is well-known that a combination of advection of moist maritime air and orographic lifting creates favorable conditions (Hanssen-Bauer and Førland, 2000).

Fig. 2.4. Local properties, such as tree specie, may dictate the local character (here color).

2.2.3 *A purely statistical view*

ESD can be used to model indirect relationships, where it is difficult to find one direct physical process (Christensen *et al.*, 2007). There must nevertheless be a real influence of the predictor on the predictand for the ESD to make sense. For instance, the temperature over a larger region can be used to model the flowering date of some flower (Bergant *et al.*, 2002).

2.3 What is "Skillful Scale"?

One of the earliest discussion of the issue of skillful scale can be traced to Grotch and MacCracken (1991), who compared climate sensitivity of several GCMs on a number of different spatial scales. They stated:

> "Although agreement of the [global] average is a *necessary* condition for model validation, even when averages agree *perfectly*, in practice, very large regional or pointwise differences *can*, and *do*, exist."

Grotch and MacCracken (1991) found that individual point differences in temperature can exceed 20 K. When they examined successively smaller areas, the spread between the models became more and more apparent generally. The agreements tend to be better in winter and for zonal averages than in summer and for subcontinental scales.

Grotch and MacCracken (1991) also concluded that the quality of the model simulations of present climate sets one limitation in terms of projecting a future climate change.

The paper by Grotch and MacCracken (1991) has been reinterpreted by Zorita and von Storch (1997) as: "at finer spatial resolutions, with scales of a few grid distances, climate models have much smaller skill," and by von Storch *et al.* (1993) as: "the *minimum scale* is defined as the distance between two neighboring grid points, whereas the *skillful scale* is larger than N gridpoint distances. It is likely that $N \geq 8$." The presence of a skillful scale was also acknowledged by Huth and Kyselý (2000), who also referred to the work by Grotch and MacCracken (1991).

From a modeling aspect, one may expect simulations not to be accurate on the minimum scale due to numerical noise (digitalization cannot give a perfect description of a continuous variable, numerical schemes are imperfect, etc.), approximation of unresolved processes, the necessity to describe the smallest wavelengths, and advection[1] associated with these. The concept of "skillful scale" applies to both RCMs as well as GCMs.

The work by Grotch and MacCracken (1991) was based on old models without a complete ocean–atmosphere coupling, involved short nontransient integrations, with models that included only two to nine vertical layers, and where only one GCM included the full diurnal cycle of solar radiation. Since this study, the climate models have improved.

Hence, the work by Grotch and MacCracken (1991) is due for a revision, and the term "skillful scale" seems to have one original source (Grotch and MacCracken, 1991), but remains elusive, especially with respect to the state-of-the-art global climate models.

There are different aspects that possibly may affect the skillful scale: (a) the choice between spectral models (atmospheres only) or grid-point models; (b) the numerical integration schemes and discretization; and (c) the surface process parameterization schemes. For instance, one type of time integration commonly used in the GCMs is the *leapfrog* scheme, but often the numerical solutions from such algorithms contain spurious oscillations known as "numerical mode" (Satoh, 2004).

---

[1]Transport of some quantity, carried with the fluid motion.

The time integrations may be less relevant for the skillful scale than the techniques dealing with the spatial gradients and partial differential equations. The spectral models employ a Fourier transform method (Press *et al.*, 1989, p. 704) to compute the gradient (Poisson equation) rather than finite differencing. The very different nature of these algorithms is likely to affect aspects such as numerical modes, as well as the consistencies with parameterization schemes, radiative models etc.

Few studies have to date explored how the skillful scale depends on these choices. One obstacle is the difficulty in defining and estimating what the skillful scale is in the first place.

The implication of "skillful scale" is central to ESD. A statistical model applied to just one model grid box may arguably be considered as a kind of downscaling if the grid box represents an area rather than a point measurement.

One may on the other hand argue that a downscaling involves going from a skillful description of the large-scale condition to the small-scale that cannot skillfully be represented by the model. We will reserve the term ESD for the latter, and use the term "adjustment" or "assimilation" for the former.

It is also important to keep in mind the difference between *skillful scale* and the *optimal domain* which will be discussed later on. The former is the smallest spatial scale for which the GCMs is able to provide an adequate representation, while the optimal domain refers to the size of the area represented by the large-scale predictor, that is greater than the skillful scale, yielding the representation with maximum correlation with the predictand.

The skillful scale may be defined differently depending on what aspect of skill is required, e.g. the mean value, the climatology, the variance, or the time structure (autocorrelation). One criterion may be that the model produces the same probability distribution (PDF) as the empirical data for the control period.

Figure 2.5 shows the comparison between PDFs drawn from a GCM, RCM, and ESD and compare these with the actual distribution derived from observations. It is immediately apparent from the comparison of the PDFs for the daily mean temperature, that the GCM does not give similar statistics on the grid-box level.

The RCM with a $50° \times 50°$ spatial resolution provides an even worse description of the distribution in this case. However, the RCM and GCM

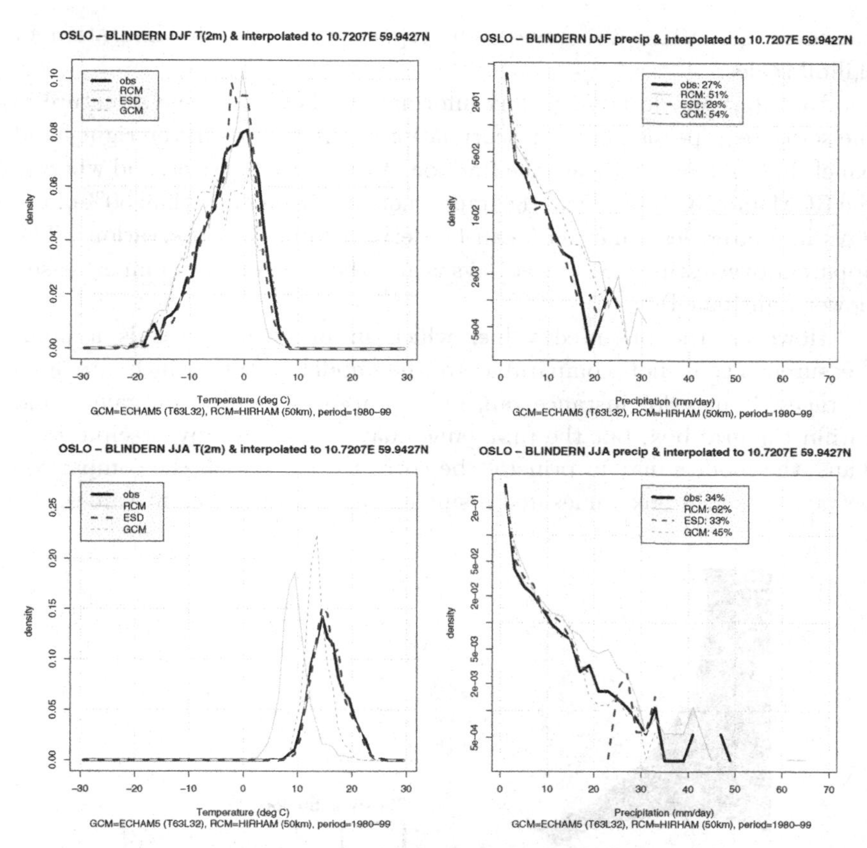

Fig. 2.5. Example of distributions of grid-box values from a GCM (interpolated), an RCM (interpolated) and ESD, as well as corresponding observations. The upper panels show the PDF for winter daily mean temperature (left) and 24-h precipitation (right), all for the period 1981–2000. Lower panels show corresponding results for the summer. Note, the vertical axes for precipitation have a log-scale.

shown in Fig. 2.5 are not directly comparable, as the RCM is driven by a different (older) model than the GCM (which is from the IPCC fourth assessment simulations). The main point here is that neither the GCM nor the RCM provide a good description for the local climate in general (if they do, it is accidental).

The ESD results (Fig. 2.5) provide a closer description of the PDF, which is expected since the models are tuned to these data. The comparison between predictions of the local 24-h summer-time precipitation and measured quantities also reveals discrepancies due to the limitation of skillful scales.

Distributions do not give any information about the time structure in the series (e.g. persistence, autocorrelation). The legends in the right-hand panels in Fig. 2.5 do indicate the fraction of wet days, however, and whereas the RCM and GCM may give the impression of rain on more than 50% of the days in winter, less than 30% can be derived from the observations. The apparent over-estimation of wet days is also the case for the summer season (lower right panel).

However, the observed value, which in practical terms is a point measurement, is not comparable to the model results which are area averages. It may for instance rain in a location near the rain gauge, and within the grid box, but the rain gauge may not record any precipitation. Thus, the models may in principle be correct even though the comparison between the grid box values and a single station show different values.

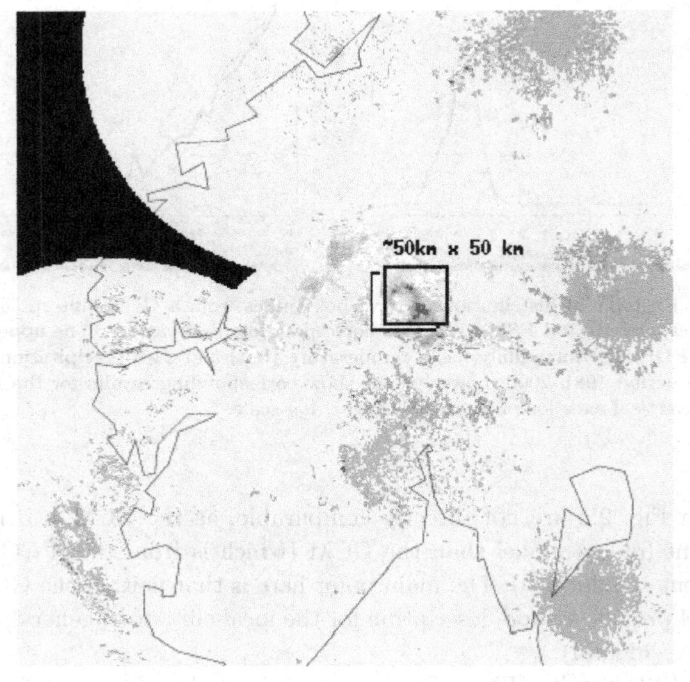

Fig. 2.6. An example showing the spatial structure of precipitation from radar reflection and a typical size of an RCM grid box, showing spatial variations at scales smaller than the models spatial resolution.

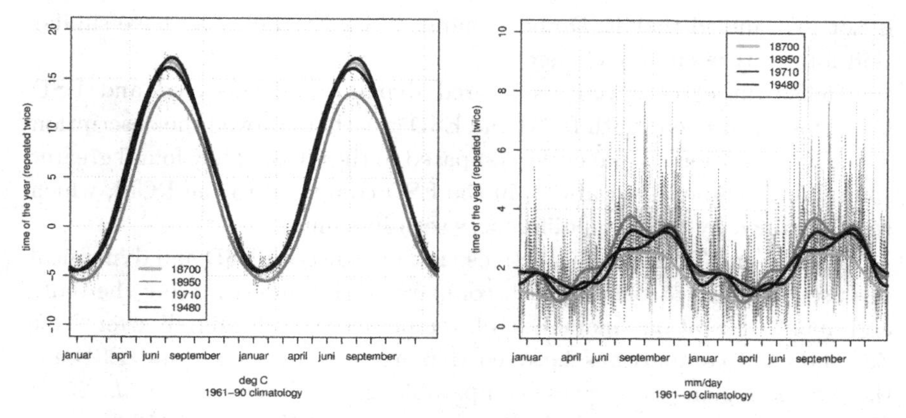

Fig. 2.7. Example of distributions of how the climatology may differ at nearby locations separated by distances smaller than the minimum scale.

Figure 2.6 gives a picture of how precipitation varies on small spatial scales compared to a typical grid-box area, and Fig. 2.7 shows how the climatology may differ for nearby stations, even within the distance of the minimum scale.

Is it possible to use observations to evaluate models then, if the models cannot represent the local scale at which the empirical measurements are made? The best practice is to aggregate (sum) observations in order to provide regional mean values, rather than trying to focus on smaller scales.

The model representation of meteorological/climatic parameters is characterized by overly spatially smooth fields/maps, but by aggregating regional climatic information, it is possible to arrive at corresponding smooth quantities.

2.4 Further Reading

2.4.1 *Dynamical versus empirical–statistical downscaling*

Murphy (1999) argued that dynamical downscaling and ESD show similar skill, but that ESD was better for summer-time estimates of temperature and dynamical methods yielded slightly better estimates of winter-time precipitation. He argued that the skill with which the present-day surface climate anomalies can be derived from atmospheric observations is not improved using sophisticated calculations of subgrid-scale processes made in climate models rather than simple empirical relations. Furthermore, it is not guaranteed that ESD and dynamical downscaling may have similar skill for a future climate change.

Hellström *et al.* (2001) compared dynamical downscaled and ESD scenarios for Sweden. Both RCM and ESD greatly improved the description of the precipitation annual cycle, compared to the GCM. They found greater spatial and temporal variability in the ESD compared to the RCM, which was explained by the large differences seen in summer.

Kidson and Thompson (1998) also found that both ESD and dynamical downscaling were associated with comparable skill in estimating the daily and monthly station anomalies of temperature and rainfall over New Zealand, and comparisons between dynamical downscaling and ESD for the Nordic region also suggest comparable skill (Christensen *et al.*, 2007; Houghton *et al.*, 2001; Hanssen-Bauer *et al.*, 2003; Kaas *et al.*, 1998).

Dynamical downscaling and ESD should support each other (Oshima *et al.*, 2002). Murphy (2000) have argued that the confidence in estimates of regional climate change will only be improved by the convergence between dynamical and statistical predictions or by the emergence of clear evidence supporting the use of a single preferred method.

Although statistical downscaling may have similar merits as dynamical downscaling (Christensen *et al.*, 2007; Houghton *et al.*, 2001; Kidson and Thompson, 1998; Kaas *et al.*, 1998), the different methods have different strengths and weaknesses. The statistical downscaling method tends to give a greater geographical spread in the mean temperatures than scenarios using the predictions from GCM grid boxes (Hulme and Jenkins, 1998) or the dynamical projections (Räisänen *et al.*, 1999; Rummukainen *et al.*, 1998).

Statistical downscaling, on the other hand, may be insensitive to some of the systematic biases if the GCM results are projected onto the "correct" observed climate patterns. However, the statistical models can also give misleading results if the modeled spatial patterns projects onto "wrong" observed patterns.

It is important to realize that the issue of stationarity not only applies to ESD, but to GCMs and dynamical downscaling as well. The dynamical climate models all involve different types of sub-grid parameterization schemes, which are just statistical models similar to ESD. Parameterization schemes tuned for the present-day conditions are not guaranteed to be valid in a future climatic state. Nonstationarity in the parameterization schemes may indeed be more serious than in ESD (thus provide a "slippery slope"), as these calculations are feed back into the model and can have more dramatic effect through many iterations.

One common misconception is the notion that RCMs provide a physically consistent description of the regional climate. Although the equations provide a dynamical solution based on the Navier–Stokes equation, the numerical methods solving these are often imperfect due to discretization. Furthermore, the parameterization schemes provide approximate bulk formula descriptions of subprocesses and land surface processes which may not give exact representations. There may also be difficulties associated with lateral boundary conditions (Kao *et al.*, 1998; Mahadevan and Archer, 1998), and schemes to inhibit overly strong convection or filter out unphysical wave solutions artificially constrain the model solutions to a state that is close to the real world, but implies a lack of physical consistency.

In summary, both downscaling approaches have their weaknesses and strengths, and it is difficult to say which is superior. It is clear that different methods are appropriate for different use, and it is therefore important to apply both dynamical and statistical downscaling to global GHG GCM results. A comparison between the two fundamentally different techniques can give us a measure of uncertainty associated with the predictions. If both methods give similar answers, then we may at least have *some* confidence in the results.

2.5 Examples

2.5.1 *Geostrophic balance*

An example of inferring one scale from another is to use the equation for geostrophic balance and pressure measurements from two locations to model the local wind in a point bisecting the line between the two pressure measurements (here in this example, only one component of the wind will be estimated — the component perpendicular to the line between the p measurements).

The local wind measurement represents the small-scale (affected by turbulence), whereas the p measurements and the geostrophic frame work represents the large scale. It is assumed that the local wind is affected by the large-scale flow between the points.

2.5.2 *Basic preprocessing*

Much of the work in ESD involves preparing (preprocessing) data so that the computer code can analyze them. The predictand and the predictors must be synchronized, so that the same times for the local scale and the large scales are linked.

Some aspects of the preprocessing are discussed in the next chapters, but we will go through some of the more basic ideas here. ESD always starts by retrieving (or reading) in the data to the computer memory. Usually the data are read from disc, but can also be read over the Internet.

In R there are various ways of reading the data, and the data are represented in entities referred to as "objects" (a bit analogous to variables in other computer languages, but may have a complex structure).

The clim.pact-package comes with some sample data which can be retrieved with the data() command:

```
> library(clim.pact)        # Activate clim.pact
> # Example of reading in a predictand:
> # Retrieve the monthly mean T(2m) for Oslo:
> data(oslo.t2m)
> # Information about this data can be found:
> ? oslo.t2m
> # Retrieve the daily mean T(2m) and precip for Oslo:
> data(oslo.dm)
> class(oslo.t2m)
> class(oslo.dm)
> summary(oslo.t2m)
> summary(oslo.dm)
```

The symbol "#" is used for commenting in R, and "?" gives on-line help information on functions and data in the installed packages.

In the above example, the monthly mean T(2m) for Oslo is first read into memory, and then a query is made about the data. The following line reads daily mean temperature and precipitation for Oslo. Note the suffix "'.dm" denotes "daily mean" in clim.pact.

The command class() shows what type of object the two examples here are, and this information is used by the functions to decide how to treat the objects. The last two lines in this example employ the summary()-command, which can be used to see the contents of the objects.

Note, the daily mean station objects look different to the monthly objects, as the former holds two parameters (typically "t2m" and "precip") while the latter is designed to hold one parameter ("val" for "value").

The data()-command can only be used to retrieve data already incorporated in the installed packages. In clim.pact, there are also other functions to read in data:

```
> library(clim.pact)        # Activate clim.pact
> obs1 <- getnordklim("oslo") # Read data from the Nordklim project
> #(http://www.smhi.se/hfa_coord/nordklim/)
> obs2 <- getnacd("oslo")     # Read data from the NACD project
> obs3 <- getnarp("Tromsoe")  # Read data from the NARP project
> obs4 <- getgiss()           # Read data from GISS.
> obs5 <- getecsn()           # Read data from the ECSN project.
> # Check the results for obs1
> plotStation(obs1)
```

All these calls returns a monthly station object that clim.pact knows how to deal with. For instance, the clim.pact-function plotStation() is designed to make a graphical presentation of these objects.

The getnordklim and getnacd calls require that the data are already installed on the computer in the NACD format (Frich *et al.*, 1996).

It is also possible to read the data as an ordinary table from a text file, and then create a station object:

```
> library(clim.pact)
># Read in the Monthly England & Wales precipitation (mm) over
the Internet.
> a <- read.table("http://www.metoffice.gov.uk/research/hadleycentre
/CR_data/Monthly/HadEWP_act.txt",skip=3,header=TRUE)
> class(a)
> # Data preparation: store the acual measurements in a matrix
called 'X'
> X <- cbind(a$JAN,a$FEB,a$MAR,a$APR,a$MAY,a$JUN,a$JUL,a$AUG,a$SEP,
a$OCT,a$NOV,a$DEC)
> # Inspect the results:
> class(X)
> dim(X)
> # Transform the data to a 'station object'
> precip <- station.obj(x=X,yy=a$YEAR$,obs.name="Monthly England &
Wales precipitation",unit="mm"ele=601)
> # The pre-processing is complete - now check the results!
> plotStation(precip)
> ? station.obj
```

Although the monthly England and Wales precipitation is strictly not a station record, we can treat it as if it were in the analysis here.

2.6 Exercises

1. Which four conditions need to be fulfilled for ESD? Why are these important?
2. What are the two main approaches for downscaling? How do they differ?
3. List the caveats of ESD.
4. What kinds of shortcomings may be a problem for dynamical downscaling?
5. Why is it important to apply both ESD and dynamical downscaling?
6. What is meant by "skillful scale"? How does this differ from "minimum scale"?
7. Start `R` and install clim.pact. In the Windows version, the installation of packages is easy (tools on top bar). In Linux, you will have to download the `clim.pact`-package from CRAN, set a system variable `R_LIBS` (create a local directory, and set `R_LIBS` to this location), and run a command in a Linux shell: "`# R CMD INSTALL clim.pact clim.pact_2.2-5.tar.gz`." The installation depends on the `ncdf` and `akima` (available from CRAN), so these must be installed prior to `clim.pact`. Do the examples above. (Some of the calls may not work, e.g. `getnordklim`, `getnacd`, `getgiss`, `getecsn`.)
8. Read a data series (own or over the Internet). Use the example above, and make your own station object. Use plotStation to make a graphical visualization.

Chapter 3

PREDICTORS AND PREPROCESSING

Several large-scale climate variables have been used as predictors of the statistical models. Due to its strong influences on the local climate (e.g., Chen and Hellström, 1999), atmospheric circulation is usually the first candidate of predictors. Among various ways to characterize the circulation, indices (e.g. Chen, 2000) and some kind of EOF analysis based on air pressure and/or geopotential height data are often used (Benestad, 2001a).

In this chapter, we look at different ways to prepare the predictor data before they can be used as inputs in statistical modeling. We start from the simple approaches based on "circulation indices," and proceed to EOF analysis for multivariate variables, and how to apply EOF analysis to reduce the problem of colinearity.

3.1 Choosing Predictors

Large-scale climate variables other than standard climate indices have been used in ESD. For example, Kaas and Frich (1995) stated that the inclusion of tropospheric temperature information among the predictors is of fundamental importance for estimating greenhouse gas induced changes. They thus used both the 500–1000 hPa thickness and the sea level pressure (SLP) fields as predictors.

Several potential "signal-bearing" predictors have been tested for downscaling precipitation. Hellström *et al.* (2001) used large-scale absolute humidity at 850 hPa (q850) as predictor for precipitation, in addition to circulation indices. They conclude that changes in q850 seem to convey much of the information on precipitation changes projected by ECHAM4.

Fig. 3.1. Past traditions and myths have given rowan berries the dubious ability to predict the subsequent winter's snows.

Linderson *et al.* (2004) tested several predictors for monthly mean precipitation and frequency of wet days, including large-scale precipitation, humidity and temperature at 850 hPa, and the thermal stability index. They concluded that large-scale precipitation and relative humidity at 850 hPa were the most useful predictors in addition to the SLP-based circulation indices. Relative humidity was more important than precipitation for downscaling frequency of wet days, while large-scale precipitation was more important for downscaling precipitation.

3.2 Circulation Indices

The most used ciculation indices are probably the Southern Oscillation Index (SOI) and related El Niño indices (Philander, 1989), the North Atlantic Oscillation Index (NAOI), and the Arctic Oscillation Index (AOI). Figure 3.2 provide an illustration of the chronological variations in the NAO and Nino3.4 indices (mean SST anomaly over the 5°S–5°N/120°W–170°W region).

These give a simple description of the situation over a large region. In addition, it is possible to use other data, such as the sunspot number,

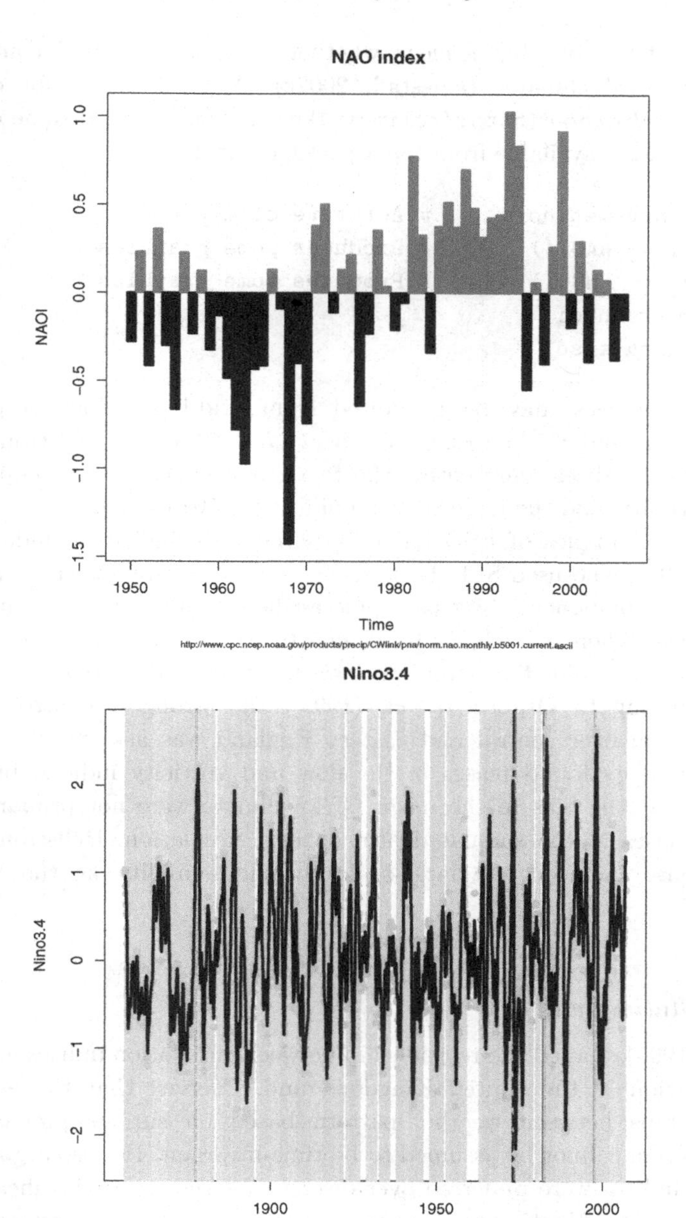

Fig. 3.2. The graphics produced by the example: top the NAOI and bottom the NINO3.4 index.

although it is not always clear whether these have a real connection with the local climate (Benestad, 2002c). A good source for data is the *ClimateExplorer* (`http://climexp.knmi.nl/`), however, some climate indices are also available from the R-package `met.no`.

```
> library(met.no)        # Activate clim.pact
> naoi <- NAOI()         # Produces some graphics
> enso <- ENSO()         # Produces some graphics
> summary(naoi)
> summary(enso)
```

Some indices may be generated from grid-box values of gridded observations and GCM results. Chen *et al.* (2005) and Hellström *et al.* (2001) used indices representing the two geostrophical wind components, total vorticity, and the large-scale humidity at 850 hPa height.

Other examples of ESD based on circulation indices include Kilsby *et al.* (1998), who used SLP, total shear vorticity as well as the zonal and meridional components of geostrophic air-flow estimated for a number of grid-points, whereas both Wilby *et al.* (1998) and Osborne *et al.* (1999) used the geostrophic flow strength, flow direction, and vorticity.

According to Osborne *et al.* (1999), the strongest control on the precipitation over central and eastern England was associated with the vorticity. They found biases in the flow and vorticity indices, but they argued that the bias in the mean GCM climate were not primarily due to the biases in the simulated atmospheric circulation. Hellström *et al.* (2001) also included the large-scale specific humidity at the 850 hPa level.

3.2.1 *Stationarity*

Wilby (1997) studied the relationship between circulation indices and local precipitation in the United Kingdom, and observed that the empirical relationships between weather–pattern based on surface pressure and precipitation cannot be assumed to be time-invariant. He also argued that air-flow indices were preferred over discrete weather type classification as a means of investigating nonstationary relationships for precipitation over time, as the former involve continuous variables that do not impose artificial boundaries on the data set and do not restrict the sample sizes to the same extent.

3.2.2 *The North Atlantic Oscillation (NAO)*

The North Atlantic Oscillation (NAO) is important for the climate in the Nordic countries (Benestad, 1998a, 2001a), but it is difficult to reproduce the exact evolution in the climate models. There may be several reasons why the NAO is so difficult to predict, some of which may be due to a chaotic nature, model misrepresentation associated ocean dynamics (too low resolution), ocean–atmosphere coupling, sea-ice, and topography.

Hanssen-Bauer (1999) found that the recent climatic trends in Norway can primarily be explained in terms of systematic changes in large-scale circulation.

The recent observed winter time warming may be related to the strengthening of the winter time NAO, as the correlation between the NAOI and south Norwegian temperatures is strong. If empirical SLP and 500 hPa models, which primarily describe the relationship between the large-scale circulation and the local temperatures, do not indicate much warming, then the warming is not due to systematic shifts in the large-scale circulation. Benestad (2001a) argued that little of the warming since 1860 over Norway could be explained in terms of systematic changes to the NAO, despite the preceeding strengthening of the NAO.

3.2.3 *Autocorrelation and degrees of freedom*

When multivariate variables, such as fields with smooth variations in space, are used as predictors, there are two important aspects to consider. First, a smooth field contains redundant information. Second, the predictor describing smooth fields consists of colinear time series. The former property allows us to reduce the data size and still retain the information, while the latter limits the choice of numerical techniques that can be employed.

Spatial coherence

If there is spatial coherence in a field, i.e. that different measurements $x_r(t)$ are correlated over locations r at times t (spatial autocorrelation), then the actual number of independent spatial observations is smaller than the number of observers. The data can then be represented by matrix X with n_r simultaneous observations at different sites (grid points on a mesh), each made n_t times.

Spatial correlation relates to how smooth the field varies spatially, and it was stated earlier that spatially smooth fields are a requisite for ESD. Thus, by measuring x at r_1 we can get an idea of what the value is at the nearby location r_2. Thus, the actual degrees of freedom for each time is less than n_r (for gridded data $n_r = n_x \times n_y$). This spatial smoothness also is utilized by Empirical Orthogonal Functions (EOFs), which extract the essential information from the data matrices, since nonzero correlation implies a degree of redundancy.

However, we often assume that the data are uncorrelated in time in our analysis, i.e. X consists of independent temporal realizations. Hence, a principal component analysis (PCA) may represent the data in terms of a small number of EOFs describing the coherent spatial structures with similar "behavior."

3.2.4 *Empirical orthogonal functions*

In geosciences, gridded data fields can be thought of as a series (or stack) of maps, one for each time of observation, and often, one particular feature is prominent in several of these maps. If this feature is removed from the data by subtraction, then we will end up with a new set of data with a different character.

The particular feature that we removed can also be represented in the form of a map. However, the original data can still be recovered by adding the feature back to the new set of maps. Thus, each map can be thought of as the result of a weighted sum of a number of set of different maps (Fig. 3.3).

We call this way of splitting up one map to several others "decomposition," and each map can be represented mathematically by a vector $\vec{x}$:

$$\vec{x} = \vec{e}_1 + \vec{e}_2 + \vec{e}_3 + \cdots . \tag{3.1}$$

This concept can also be shown graphically (Fig. 3.3). But why is it interesting to decompose one map into a number of others?

It turns out that given a set of *basis maps* and giving each a different weight, then it is possible to express all other maps with a much smaller set of basis maps:

$$\vec{x}(t) = \beta_1(t)\vec{e}_1 + \beta_2(t)\vec{e}_2 + \beta_3(t)\vec{e}_3 + \beta_4(t)\vec{e}_4 + \cdots . \tag{3.2}$$

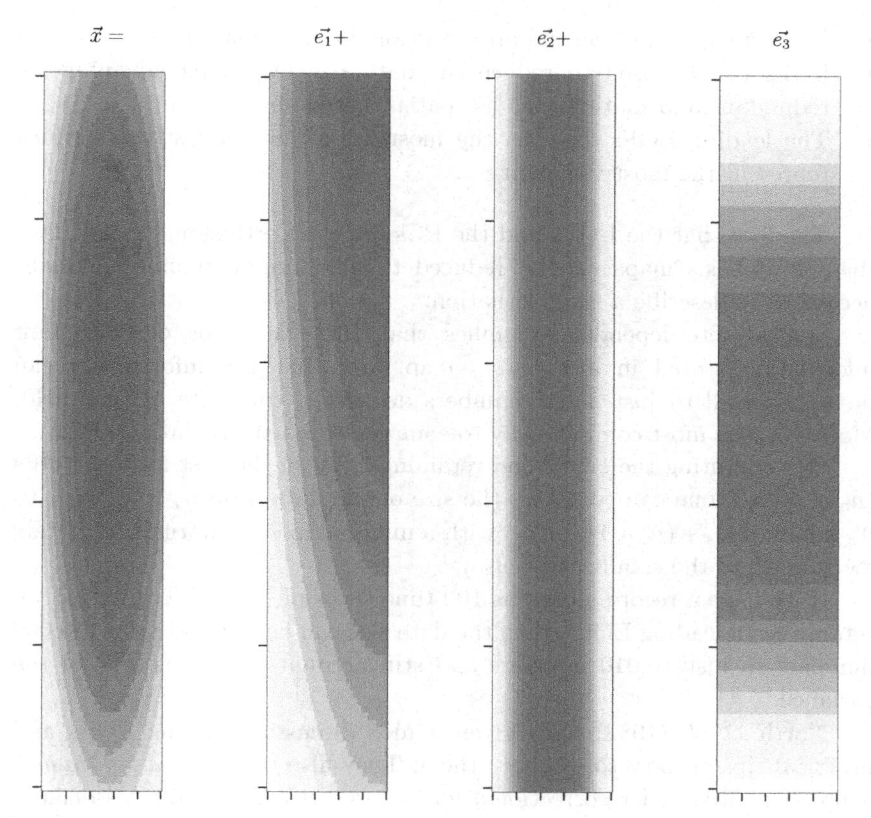

Fig. 3.3. Example showing how one map ($\vec{x}$ on the left) can be expressed as a superposition of other maps/structures ($\vec{e_1} + \vec{e_2} + \vec{e_3}$).

EOFs (Preisendorfer, 1988; North *et al.*, 1982; Lorenz, 1956) are convenient mathematical constructions which enable the identification of a small set of basis maps. They are a special product of a more general Principal Component Analysis (PCA) (Strang, 1988; Press *et al.*, 1989), but taylored for geophysical needs.

EOFs are associated with corresponding principal component (PC), also referred to as "loading vector." Whereas the EOF represents a spatial structure (a map) and is represented by $\vec{e_i}$ in Eq. (3.2), the PCs describe how strongly these are present at the time $\beta_i(t)$. Some important properties of EOFs are

- The EOFs are orthogonal (uncorrelated, or are perpendicular in "data space").

- The principal components are orthogonal (uncorrelated).
- EOFs can be used to reduce the data size by taking advantage of redundant information such as spatial correlation.
- The leading EOFs describe the most important modes; those which represent the most variability.

The fact that the EOFs and the PCs both are orthogonal means that the set of basis maps can be reduced to a minimum number of maps necessary to describe a given situation.

Spatial interdependence implies that there is a lot of redundant information stored in a $n_x \times n_y$ map, and that the information can be compressed to just a few numbers describing the state of that field. Moreover, the most common way to compress the data is through PCA.

By computing the EOFs and retaining a few of the first leading EOFs ($n_{\mathrm{eofs}} \ll n_t$), one can compress the size of the data from $n_x \times n_y \times n_t$ to $n_x \times n_y \times n_{\mathrm{eofs}} + (n_t + 1) \times n_{\mathrm{eofs}}$ with a minimal loss of information (filters away much of the small-scale noise).

If we have a record stored as 100 time slices on a 50×30 grid and we retain the 10 leading EOFs, then the data size can be reduced from 150,000 numbers to just 16,010 numbers and still account for about 90% of the variance.

North *et al.* (1982) have given a nice discussion on the EOFs and practical tips on how to estimate them. They also gave a "*rule-of-thumb*" expression (first-order corrections) for the uncertainties (shift) associated with the ith eigenvalue estimation λ_i:

$$\delta\lambda_i \approx \lambda_i \sqrt{2/N}. \tag{3.3}$$

"S-mode"

The vectors are written as $\vec{x}$ and matrices are denoted by using the capital letters: $X = [\vec{x}_1, \vec{x}_2, \ldots, \vec{x}_T]$. The vector quantities are used to represent several observations at a given time, i.e. they can be regarded as maps. Let the *number of observers* mean the number of grid points or stations where observations are made (number of observers = R), and the *number of observations* be the length of the time series at each location (number of observations = T). We use the notation $\bar{x}$ to mean the temporal mean of x and $\langle x \rangle$ the spatial (ensemble) mean of x. Let the matrix X_{rt} contain T observations from R different locations, where X can be expressed in the form $X = [\vec{x}_1, \vec{x}_2, \ldots, \vec{x}_T]$ and $\vec{x}_t = [x_1(t), x_2(t), \ldots, x_R(t)]$. Each column represents one set of observations, with each element holding the data from the R different locations:

$$X = \begin{pmatrix} \cdots & \rightarrow & T \\ \downarrow & \cdots & \cdots \\ R & \cdots & \cdots \end{pmatrix}. \tag{3.4}$$

Let anomalies in X be defined as

$$X'_{rt} = X_{rt} - \frac{1}{T}\Sigma_{t=1}^{T} X_{rt} = X_{rt} - \bar{X}_r. \tag{3.5}$$

The variance–covariance matrix is defined as

$$C_{rr} = X'X'^T = \begin{pmatrix} \cdots & \rightarrow & R \\ \downarrow & \cdots & \cdots \\ R & \cdots & \cdots \end{pmatrix}. \tag{3.6}$$

The S-mode Empirical Orthogonal Functions (EOFs) of X_{rt} are defined as

$$C_{rr}\vec{e}_s = \lambda \vec{e}_s. \tag{3.7}$$

Let $E_s = [\vec{e}_1, \vec{e}_2, \ldots, e_{\vec{R}^*}]$ be a matrix with the columns holding the eigenvectors (EOFs) and R^* be the rank of X. The original data may be expressed in terms of the orthogonal set spanned by the EOFs:

$$X = E\beta \tag{3.8}$$

where β is the projection of X onto the EOF space.

We can use singular-value decomposition (SVD) to compute the EOFs. Using SVD, we can express the matrix X' as

$$X' = U\Sigma V^{\mathrm{T}}. \tag{3.9}$$

Note that the SVD algorithm is written in such a way that the numbers of columns must be less than number of rows. In this example, the number of observers are assumed to be greater than the number of observations (which often is the case for gridded climate data). If the number of columns is greater than the number of rows, then the SVD must be applied to the transpose of the matrix (U and V will now by swapped).

The columns of U and V are orthogonal respectively:

$$U^{\mathrm{T}}U = V^{\mathrm{T}}V = I. \tag{3.10}$$

The matrix Σ is a diagonal matrix, with R^* nonzero singular values and $R-R^*$ zero values in descending order along the diagonal. The inverse of Σ is a diagonal matrix with the reciprocal of the nonzero singular values along the diagonal, and where the reciprocal of the small singular values or zeros are taken to be zero. The variance–covariance matrix can be expressed in terms of the SVD products:

$$C_{rr} = X'X'^{\mathrm{T}} = U\Sigma V^{\mathrm{T}}(U\Sigma V^{\mathrm{T}})^{\mathrm{T}} = U\Sigma V^{\mathrm{T}}(V\Sigma U^{\mathrm{T}}) = U\Sigma^2 U^{\mathrm{T}}. \tag{3.11}$$

A right operation of U gives:

$$C_{rr}U = U\Sigma^2. \tag{3.12}$$

or

$$C_{rr}\vec{u} = \sigma^2\vec{u}. \tag{3.13}$$

Hence, $U = E_s$ and $\sigma^2 = \lambda$, and the SVD routine applied to X gives the S-mode EOFs of X. The S-mode, described above, is usually employed when deriving spatial EOF maps.

Temporal coherence

If there is serial temporal correlation then the actual number of independent observations is smaller than n_t. The EOFs hence yield a smaller set of temporal structures, or "trajectories." Each of these trajectory is associated with a spatial structure given by $\beta = E^{\mathrm{T}}X$.

Spatial anomalies

We have so far only considered anomalies where the temporal mean value at each location is subtracted from the respective time series. It is also possible to perform EOF analysis on "spatial anomalies" where the mean observation at time t, $\langle\vec{x}(t)\rangle$, is subtracted from all observations at this time:

$$X^+_{rt} = X_{rt} - \Sigma_{r=1}^{R} X_{rt} = X_{rt} - \langle X_t\rangle. \tag{3.14}$$

Whereas the temporal (the usual definition of) anomalies captures trends in time (such as a global warming) and oscillations, EOF analysis based on spatial anomalies will be insensitive to the evolution of global mean values. The PCA on spatial anomalies, on the other hand, will be sensitive to large spatial gradients, although oscillating structures that have sufficiently small scales to produce large spatial variance will also be captured by the spatial anomaly EOFs.

"T-mode"

The spatial variance–covariance matrix is defined as

$$C_{tt} = X'^{\mathrm{T}} X' = \begin{pmatrix} \cdots & \rightarrow & T \\ \downarrow & \cdots & \cdots \\ T & \cdots & \cdots \end{pmatrix}. \tag{3.15}$$

The T-mode Empirical Orthogonal Functions (EOFs) of X_{rt} are defined as

$$C_{tt}\vec{e}_t = \lambda \vec{e}_t. \tag{3.16}$$

The spatial variance–covariance matrix can be expressed in therms of the SVD products:

$$C_{tt} = X'^{\mathrm{T}} X' = (U\Sigma V^{\mathrm{T}})^{\mathrm{T}} U\Sigma V^{\mathrm{T}} = (V\Sigma U^{\mathrm{T}})U\Sigma V^{\mathrm{T}} = V\Sigma^2 V^{\mathrm{T}}. \tag{3.17}$$

A right operation of V gives:

$$C_{tt} V = V\Sigma^2. \tag{3.18}$$

Hence, $V = E_t$ and $\sigma^2 = \lambda$, and the SVD routine applied to X also gives the T-mode EOFs of X.

The T-mode has been employed where temporal evolution of coherent spatial structures have been discussed. The T-mode forms the basis for both canonical correlation analysis (CCA) and regression. Note, the SVD algorithm yields both S and T mode EOFs, where S-modes are the usual maps and T-modes often are referred to as principal components (PCs).

The n_r number of independent realizations in X is often smaller than the (effective) time dimension. Therefore, the estimation of the spatial variance–covariance matrix tends to be associated with large sampling errors. In this case, the S-mode is preferred method.

The n_t number of independent realizations in T is often smaller than the (effective) spatial dimension. Therefore, the estimation of the variance–covariance matrix tends to be associated with large sampling errors. In this case, the T-mode is preferred method.

Analogies

The EOF analysis may be thought of as being analogous to data reconstruction based on Fourier transforms (FT), in the sense that both produce series (vectors) which form an orthogonal basis. The transform $f(x,t) \rightarrow F(k,\omega)$, whereby the inverse transform for each of the wave numbers k_i give sinusoidal functions which are normal to the functions of other wave numbers.

Another way of thinking about EOFs is that of a rotation in data space — a bit analogous to rotating an object when you look at it. The information is there regardless from which angle you look at, but the rotation determines which feature is most visible. We can illustrate this principle through a two-dimensional (2D) data set (bivariate data) in Fig. 3.4. The conventional axes are the horizontal and the vertical (in this case, they are orthogonal).

Since the scatter of points clearly follows a line, the two data are not completely unrelated, but one contains some information about the other. If we now rotate the whole constellation of points so that the line (red line in Fig. 3.4) on which the point lie follow the x-axis, then x will describe most of the variance in the combined data. The vector describing the new x-axis is then the leading EOF.

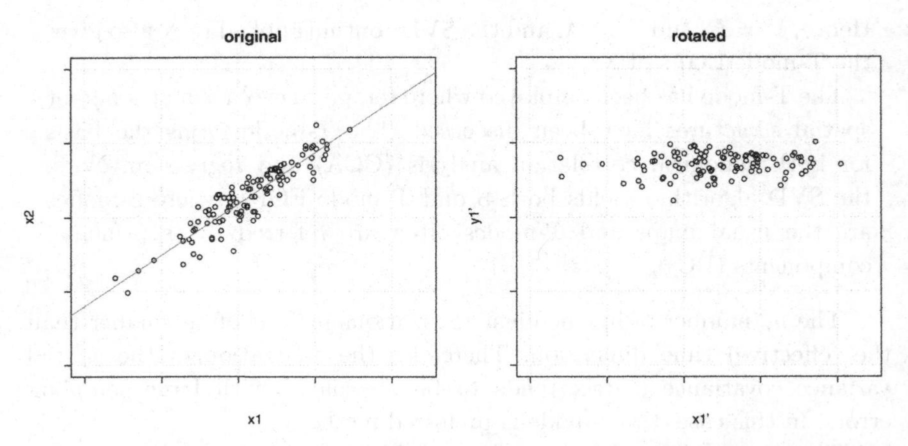

Fig. 3.4. A 2D example of the rotation of the reference axes in data space. The red line shows the direction along which the scatter is greatest.

EOF analysis is similar to eigenvalue analysis in the sense of identifying the direction in which the data exhibits the greatest scatter. Thus, the EOFs can be regarded as a kind of eigenvectors, which are aligned so that the leading EOFs describe the spatially coherent pattern that maximizes its variance. The EOFs are often used as basis functions (a new set of axes or reference frame).

Geographical weighting

It is important to apply a geographical weighting factor if the data are represented on grids that cover large latitudinal ranges, as the boxes (on a regular lon–lat grid) near the poles tend to represent a much smaller area than those near the equator.

A common spatial weighting function is $W = W_x \times W_y$, and should be applied to the data prior to the PCA, where $W_x = 1$ and $W_y = |\cos\phi|$. Then the inverse weights should be applied to the EOFs after the calculations.

Unweighted data will give too much weight to polar regions.

Similarly, for a network of unevenly distributed observers, a weighting function must be applied in order to ensure equal contribution from each *independent* data point.

Sometimes data from unwanted regions may be blanked out by setting them to zero. One reason for not removing the unwanted remote areas all together can be the desire to retain all the spatial grid points as some algorithms require more spatial data points than temporal data points. Furthermore, it may be possible to get a better estimate of the covariance matrix and hence a better estimate of the spatial patterns if these regions represent weak noise.

3.2.5 *Varieties of EOF analysis**

3.2.5.1 *Rotated EOFs**

Sometimes, the interpretation of the EOF patterns may be difficult because the adjacent modes are degenerate (not well-resolved in terms of their eigenvalues), e.g. as described in Eq. (3.3): any combination of degenerate patterns is equally valid. Furthermore, the order of degenerate modes are arbitrary. In order to resolve the modes, it is possible to rotate the EOFs.

The rotation transforms the EOFs into a *nonorthogonal* linear basis, one common type being the Varimax rotation (Kaiser, 1958) is one of the most commonly used type of rotation that minimizes the "simplicity" functional:

$$\mathbf{V}_k^* = \frac{L\sum_{i=1}^{L}\mathbf{E}_{i,k}^4 - (\sum_{i=1}^{L}\mathbf{E}_{i,k}^2)^2}{L^2},$$

V_k^* maximizes if $\mathbf{E}_{\vec{r},k}$ are all 0 or 1. $\mathbf{E}^{(R)} = \mathbf{E}\mathbf{T}^{-1}$. If two patterns are degenerate and located in different regions, rotated EOFs should resolve them. Of course, there is a catch: two waves may be degenerate.

But, what physical meaning do the EOFs actually have? Coherent spatial patterns with maximum variance. Modes of energy? Just convenient mathematical abstractions? The analysis depends on the nature of the problem.

Joliffe (2003) cautions against unjustified interpretations of rotated EOFs. He argued that it is impossible from any purely mathematical or statistical technique to find the modes in nature without any prior knowledge of their structure, and that rotation toward a simpler structure is irrelevant as the simplest structure is trivial. Here, the term "mode" refers to the spatial structure of a natural (and persisting) oscillation. Jolliffe argues that EOF analysis will be unsuccessful unless the modes are uncorrelated and are related orthogonally to the variables. However, Behera *et al.* (2003) disagree with Jolliffe.

3.2.6 *Complex and frequency-domain EOFs**

3.2.6.1 *Complex EOFs**

A complex EOF analysis (Brink and Muench, 1986; Kundu and Allen, 1976) can be applied to a two-component field when we want to look for patterns which are independent of the orientation of the axes: for instance the two-wind components $\mathbf{X} = \mathbf{U} + i\mathbf{V} \rightarrow \mathbf{C}_{XX} = \mathbf{X}^*\mathbf{X}$ (here $\mathbf{X}^*$ means the complex conjugate). $\mathbf{C}_{XX}$ is a Hermitian, so the eigenvalues are real whereas the EOFs are complex.

Complex EOFs give phase information between the two components, $\tan\left(\frac{\mathrm{Im}(\mathbf{U})}{\mathrm{Re}(\mathbf{U})}\right)$, as well as their energy. Study of propagation can be based on the application of a complex EOF analysis to the same field but with a lag difference.

3.2.6.2 *Frequency-domain EOFs**

Waves have coherent structures with consistent phase relationships at different lags. Frequency-domain EOFs (Johnson and McPhaden, 1993; Wallace and Dickinson, 1972) can be used for identifying patterns associated with certain frequencies.

3.2.6.3 *Extended EOFs**

Propagation may be studied with a technique called "Extended EOFs" (EEOFs). The math is essentially the same as for ordinary EOFs, and the difference lies in the preprocessing of the data. The EEOFs maximize the variance in a $(n_x \times n_y) \times n_l$ window.

EEOFs involve computing the covariances at all spatial lags and out to time-lag $n_l - 1$.

We let $\vec{x}_i$ describe the geographically distributed data at time i, denoted by the subscript. Then $\mathbf{X} = [\vec{x}_1 \cdots \vec{x}_T] \rightarrow \mathbf{X}' = [\{\vec{x}_1 \cdots \vec{x}_L\} \cdots \{\vec{x}_{T-L} \cdots \vec{x}_T\}] = [\vec{x'}_1 \cdots \vec{x'}_{T-L}]$, where $\vec{x'}_i = \{\vec{x}_i \cdots \vec{x}_{i+L}\}$.

PCs have rank $n_t - n_l + 1$.

Advantages associated with EEOFs include: (i) more averaging $\rightarrow$ smoother patterns and sometimes better S/N; (ii) contain lag-relationship information that can help interpretation of the patterns.

Pitfalls: The eigenvectors of the inverse covariance matrix are similar to EOFs of common noise process. Thus, the errors are $\mathbf{\Omega}^{-1}\vec{e} = \vec{e}\lambda$, which can have similar solutions to the wave equation.

Sanity check: (i) compare with EEOFs applied to data filtered through a few of the leading conventional EOFs; (ii) model each PC as an AR(1) (red noise; null hypothesis) process (MC-test): H_0 = "Data consists of mutually independent, nonoscillatory processes"; (iii) Compare power in each extended EOF/PC pair with the distribution of power in that from the surrogates: if all are outside the null-distribution, H_0 can be rejected.

3.2.6.4 *Mixed-field and common EOFs**

Mixed EOFs are just like ordinary EOFs or EEOFs in mathematical sense, but differ in how the data are preprocessed and the type of data that they represent.

Instead of merging the data with a lagged version of itself, the mixed EOFs can be calculated by merging two different fields. Their construction is very much like the common-EOFs, but now the grids of two different fields are merged so that the spatial grids are affected rather than the lengths.

Mixed-EOFs describe how two different fields, such as SLP and temperature, covary in time, and represent the same as fields-combined PCA, "CPCA," discussed in Bretherton *et al.* (1992). Thus, the mixed-EOFs tend to emphasize on covarying signals in different data fields, or coupled structures.

The different data sets in mixed-field EOFs, on the other hand, can represent different physical quantities and be stored on different grids. The different data sets, however, should be weighted in mixed-EOF analysis, so that one set does not dominate over the other.

Common EOFs, which will be discussed in more detail later on in association with ESD, is similar to mixed-field EOFs, but now one data set is appendend to another. The two data sets must be stored on the same grid, and should represent the same quantity.

The common EOF method is a useful technique for extracting common spatial climate patterns in two or more data sets. The principle of the common EOF method is that two or more data fields with data points on a common grid are combined along the time axis (concatenated), and an EOF analysis (Benestad, 1999d) is applied to the results.

Figure 3.5 provides a graphical representation of the common EOFs. The common EOFs are also discussed by Barnett (1999).

3.2.7 *EOF in ESD*

One important aspect of the EOFs is that they satisfy the orthogonality criteria ($U^{\mathrm{T}}U = I$), which in practice may result in different ordering of the EOF patterns in slightly different data sets. This is especially the case when the EOFs are degenerate or close to being degenerate.

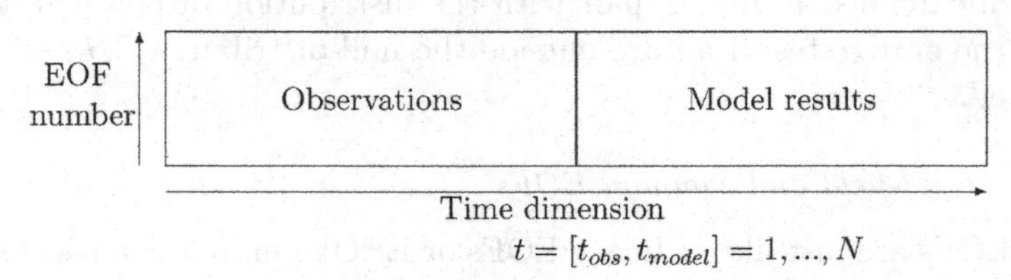

Fig. 3.5. A schematic illustration of the common EOF method, showing just the PCs. First, the PCs corresponding to the observations and station data are used for model calibration, then the corresponding PCs from the GCM model results are used for predictions.

In order to match same spatial patterns in the GCM with those found in the gridded observations, and those identified as important during ESD calibration, one can use regression to project one set of EOFs onto the other. Other techniques, such as the common EOF approach will be discussed later.

3.3 Further Reading

EOF analysis is commonly used among geophysicists, and there is a large number of references giving further details about EOF analysis and related mathematical considerations. Press *et al.* (1989) and Strang (1988) discussed the SVD algorithm in terms of numerical solutions and linear algebra, respectively. Anderson (1958) has given an account of principal component analysis from a statistical point of view on an advanced level, whereas Wilks (1995) has given a simpler introduction to EOF analysis. Preisendorfer (1988) is a commonly used text, has given detailed recipes on how to do the calculations, and Peixoto and Oort (1992) have given a brief overview of EOF analysis in one appendix. Mardia *et al.* (1979) is a good book on general multivariate methods.

Huth and Kyselý (2000) used (Varimax) rotated EOFs for downscaling monthly mean temperature and precipitation totals in the Czech republic. In order to ensure consistency between the EOFs from the observations and simulated results, they projected the observed EOFs onto the GCM results.

Huth (2004) compared results of ESD based on a various choices. He compared results based on the field directly as predictor and a number of analyses which used different numbers of EOFs/PCs and different linear methods (CCA, regression). The conclusion of his study was that temperature changes estimated though ESD depended on the number of PCs used to represent the predictors, and that the larger number of PCs, the greater the warming. The temperature change estimates varied widely among the methods as well as among the predictors.

Whereas a pointwise regression may select grid points that maximize the explained variance of the predictand, the PCs are designed to maximize the predictor variance, and necessarily contain some information irrelevant to the variability of the predictand (Huth, 2002).

Benestad *et al.* (2002) used mixed-EOFs in the downscaling of the temperature on Svalbard. The argument for using mixed-EOFs was that these may capture coupled modes, and hence be more “physical” than just ordinary EOFs.

3.4 Examples

3.4.1 *Basic EOF analysis*

In `clim.pact` there is a function for applying EOF to a data field:

```
> library(clim.pact)
> data(DNMI.slp)
> eof.1 <- EOF(DNMI.slp,mon=1)
> class(eof.1)
> summary(eof.1)
> plotEOF(eof.1)
< ? EOF
```

The function has a number of arguments to set a number of conditions which determine how the EOF is carried out. The algorithm uses the SVD method (Press *et al.*, 1989; Strang, 1988) rather than calculating the eigenvectors for a covariance matrix. By default, the function only returns the first 20 EOFs, and neglects the remaining information which in general is just noise anyway.

```
> library(clim.pact)
> data(DNMI.slp)
> eof.1 <- EOF(DNMI.t2m,mon=1)
> DNMI.slp.2 <- EOF2field(eof.1)
```

The original data can in principle be recovered completely from the EOFs, and there is a function in the `clim.pact`-package called `EOF2field()` which does exactly that. This technique can also be employed if one wants to filter out small-scale noise from the data. This filtering is simply done by applying `EOF()` to the data keeping only a few leading modes, and then do the inverse calculation by invoking `EOF2field()` (Fig. 3.6).

3.4.2 *Mixed-field EOFs*

```
> data(DNMI.sst)
> data(DNMI.slp)
> sstslp <- mixFields(DNMI.sst,DNMI.slp)
> eof.mix <- EOF(sstslp,mon=1)
> plotEOF(eof.mix)
```

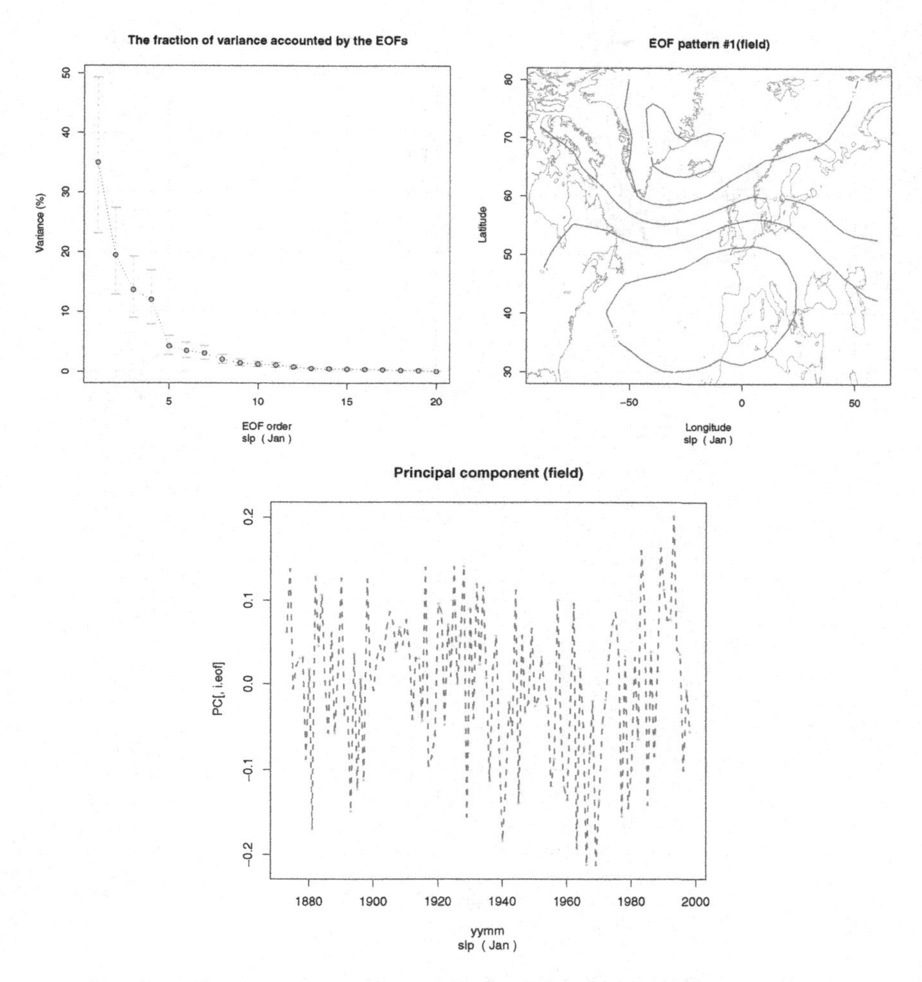

Fig. 3.6. Example of EOF results derived using the *R*-commands below.

The example in the lines above shows how easily mixed-EOFs can be constructed within the `clim.pact` framework (Fig. 3.7).

3.4.3 *Extended EOFs*

```
> data(DNMI.sst)
> eeof <- ExtEOF(DNMI.sst,mon=6)
> plotEOF(eeof)
```

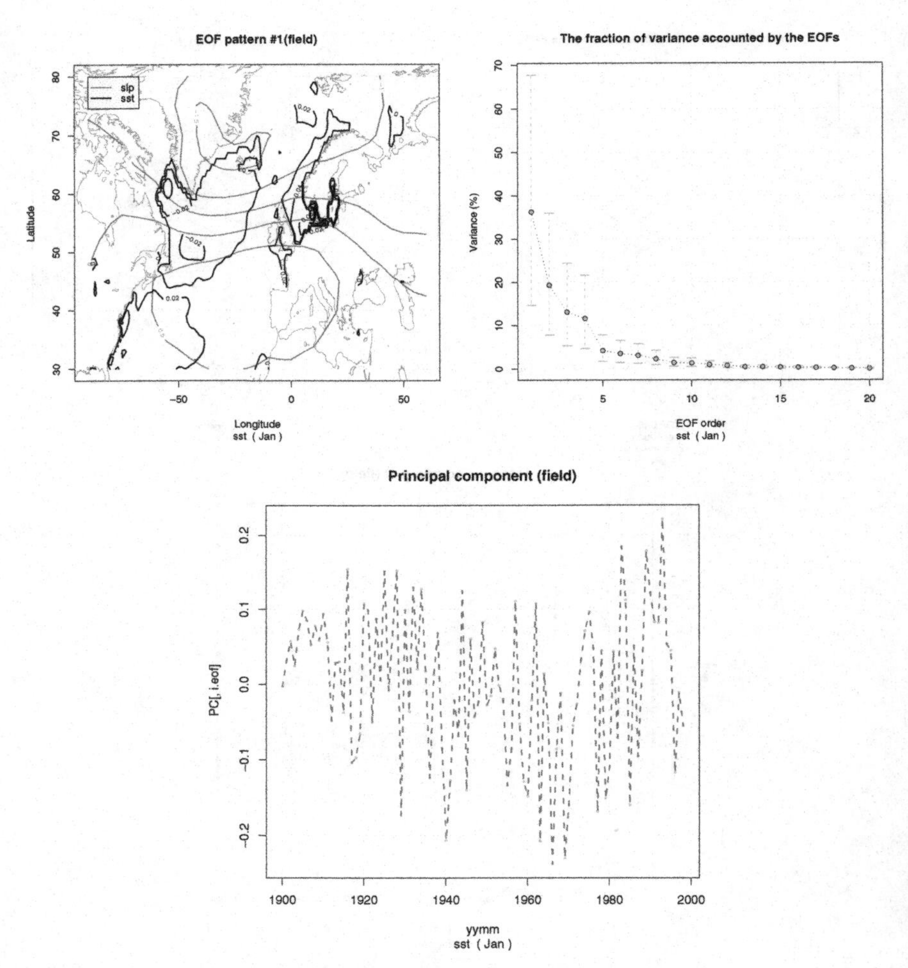

Fig. 3.7. Example of mixed-EOF output produced by the example below. The mixed-EOF describes the coupled SST–SLP January variability, exhibiting a tri-pole SST pattern associated with an NAO-type circulation structure.

The example in the lines above shows how easily extended EOFs can be constructed within the `clim.pact` framework (Fig. 3.8).

3.5 Exercises

1. Describe EOFs.
2. Estimate the EOFs of the SLP for (a) January and (b) July SLP using the period 1879–1948 from the data set *DNMI.slp*. Do the same for the

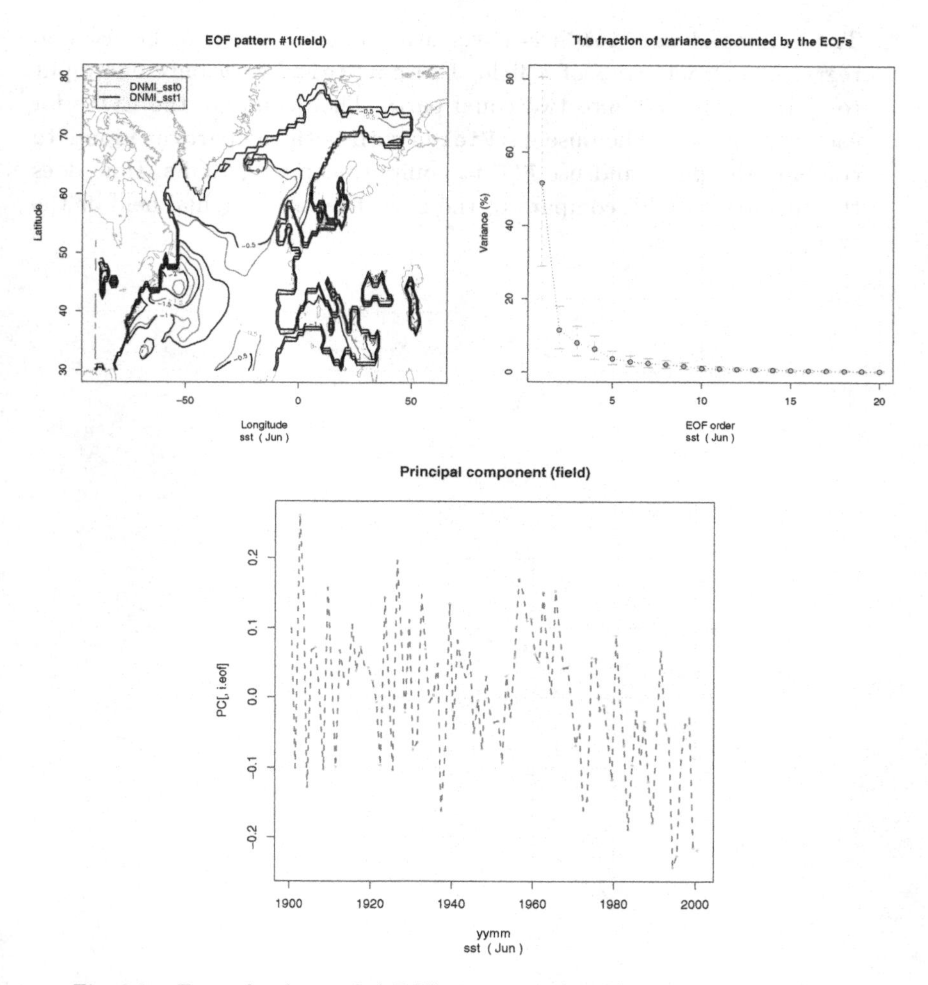

Fig. 3.8. Example of extended-EOF output produced by the example below.

period 1949–1998. Compare the eigenvalues. Are they similar? Compare the spatial patterns: are they similar?

3. Compute a set of two-component PCA for the July temperatures in Bergen and Oslo. Make (a) a scatter plot of the original data and (b) of the PCs. Compare the two plots: can you see that the PCA products give a "rotated" version of the original data?
4. Why is it useful to use EOFs and PCs in ESD rather than the fields themselves? (what useful properties do EOFs have?)
5. Use the command `mixFields` and `EOF` to compute mixed field EOFs.

6. The command `catFields` is a versatile function and can be used to regrid or extract parts of a field. Use `catFields` with only one input to split DNMI SLP into two equal parts. Then compute the EOFs for each time period. Then use `catFields` with both the parts as inputs to combine the data, and use `EOF` to compute common EOFs. How does the "common EOF" compare to the EOF from the original field or the two parts?

Chapter 4

LINEAR TECHNIQUES

4.1 Introduction

In many circumstances, climate studies involve more than just two time series. Sometimes, spatial maps of a parameter, such as the SLP, are used instead of just one station value. The geographically distributed climate quantities are often stored in *gridded* data files, consisting of $n_x \times n_y$ grid boxes (or points), each of which represent the mean value for that particular grid box area.

An example of multivariate analysis is the map of regression weight (also referred to as the "predictor pattern") between the SLP and January precipitation in Thorshavn (Fig. 4.2). Such maps can also be used to identify teleconnections, and lagged-correlation maps can be used in the study of propagating signals.

It is easy to see how linear techniques can be justified if they are viewed as a means to provide an approximate description of a response to a small perturbation (Fig. 4.3). We can divide the problem of modeling into two different classes: (a) nonlinear response to small perturbations (using the tangent of the curve) and (b) response that has a strong stochastic character. Both are common in disciplines such as physics and natural sciences.

We begin with a discussion of technical details concerning the construction of the statistical models. The first section defines linear algebra notations employed here, and is followed by sections where the equations used in Canonical Correlation Analysis (CCA) are derived. These sections are not essential for the understanding of the final results and may be skipped by those who are not interested in the model details.

Fig. 4.1. Linear techniques offer the most transparency in terms of understanding the connections, and they have one big advantage in being transparent.

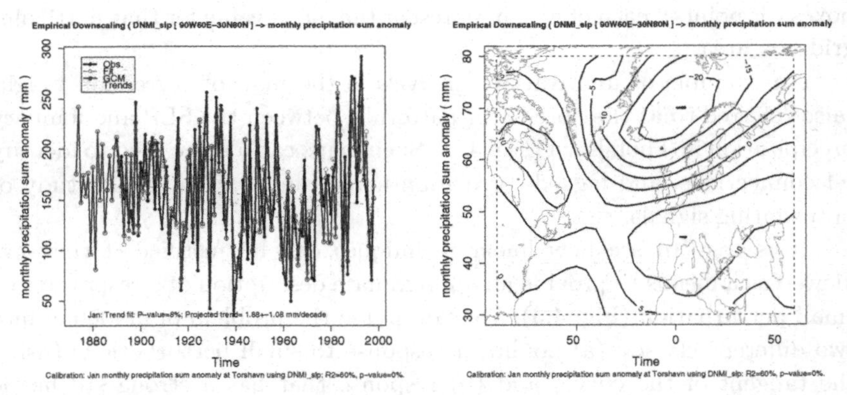

Fig. 4.2. A linear multiple regression against the January precipitation in Thorshavn and the PCs derived from the EOF analysis of the January DNMI.slp yields a multivariate analysis which return a time series predicted from the SLP data (red curve in left-hand panel) and maps of regression weight (right).

The section on cross-validation discusses the model results, i.e. the relationship between the predictor fields and the predictands. In this section, the model skill is evaluated. Where possible, a physical explanation is given as to how the predictors may influence the predictands.

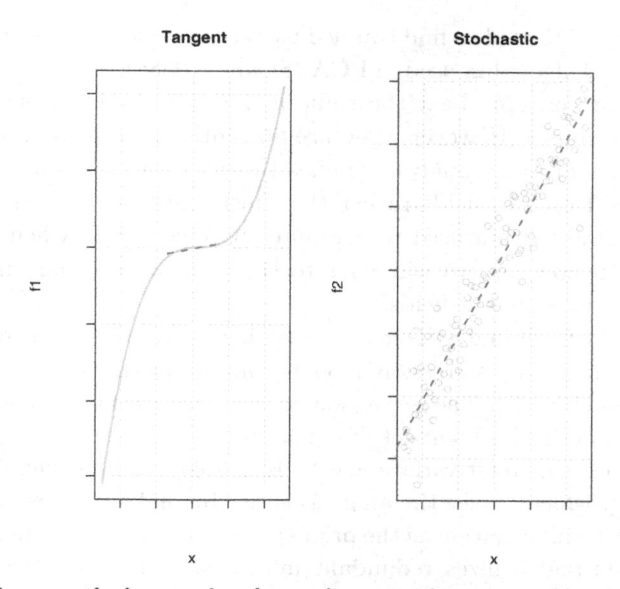

Fig. 4.3. Linear methods can often be used to provide a reasonable approximation, even to a nonlinear response as long as the perturbation is small and the response is smooth (left). Also when the response is not entirely determined by the predictors, a linear approximation may be used to provide a crude and approximate description of the response (right). The linear prediction is shown as red-dashed lines.

The section on model stationarity describes studies where the models have been constructed using one half of the data, which approximately corresponds to the periods with lowest temperatures in the northern hemisphere, and subsequently used for prediction of the second half, which is associated with warmer temperatures. This analysis is a crude test to investigate if the assumption of constant relationship between predictors and predictands holds for a warming scenario.

If the statistical downscaling models are to be used in the study of future climate change, it is important that the statistical relationship found for the training period also holds for the prediction period. The main findings are summarized in the last section.

4.2 Linear Multivariate Methods

In the category linear multivariate methods, we have three main types: Multivariate regression (MVR), Canonical correlation analysis (CCA), and Singular vector decomposition (SVD).

Here, "SVD" used to find coupled patterns should not be confused with the numerical algorithm used in PCA (Strang, 1988; Press *et al.*, 1989); the two different concepts bear the same name and abbreviations, but have different meanings. However, they are not entirely unrelated, as the SVD used in coupled mode analysis applies the SVD in PCA on a covariance matrix based on two fields rather than just one field. We will not dwell on the SVD algorithm used to compute PCA here, and when we discuss "SVD" from now on, we will refer to the method for analyzing coupled patterns in two different fields.

While the CCA and SVD models were based on maximizing correlations and covariance, the regression models aim to minimize the root-mean-square-error (RMSE). The regression models are based on the least squares solutions to an inconsistent set of equations.

Many of the multivariate methods employ a preparing step (pre-processing), which makes the analysis easier by making a new set of series with the same information as the original data, but which are now mutually uncorrelated and utilizes redundant information in such a way that the number of series can be reduced. This step involves EOFs discussed in the previous chapter.

One important part of the statistical models is that they are based on historical (empirical) data, and it is therefore crucial that the data sets on which these models are based are free from serious errors. A thorough evaluation of several predictor and predictand data sets is recommended.

The leading spatial weighting pattern (e.g. regression weights, leading spatial CCA or SVD pattern) will hereafter also be referred to simply as "the predictor pattern."

4.3 Regression

Multivariate regression (MVR) models bear many similarities with the Canonical Correlation Analysis (CCA) and Singular Vector Decomposition (SVD) techniques. MVR differs from CCA by how the model tries to optimize the fit. Whereas CCA finds pattern pairs with the highest correlation, MVR tries to minimize the root-mean-square-error.

The results from the MVR predictions are represented slightly differently here compared to the CCA and SVD models, due to the different mathematical formulations of the models.

Here, each predictand time series (station) is associated with different predictor patterns, each representing the regression coefficients between the predictor fields (e.g. 20 leading EOFs) and the respective temperature record for the particular station.

For the CCA and SVD methods, on the other hand, each predictand series may be associated with a combination of several predictor patterns. Furthermore, the predictand weights (regression coefficients of $X = \hat{Y}\Psi^{-1}$) for the MVR results indicate the contribution of each EOF to the predictor pattern, whereas for CCA and SVD, the predictand weights indicate the contribution of each CCA or SVD pattern.

4.3.1 *Multivariate regression models*

$$Y^{\mathrm{T}} = X^{\mathrm{T}}\Psi - \zeta, \tag{4.1}$$

of n equations with q unknowns, where ζ represents a noise term. Y and X have the same number of temporal samples, t, but may have different spatial representations (different grid or number of locations). Here we use the convention with spatial dimensions along the columns and temporal conventions along the rows:

$$X = \begin{pmatrix} \cdots & \rightarrow & t \\ \downarrow & \cdots & \cdots \\ r & \cdots & \cdots \end{pmatrix}. \tag{4.2}$$

Strang (1988) uses a different convention ("$Ax = b$" where "A" corresponds to X^{T} here, "x" to Ψ, and "b" to Y^{T}), but the equations are in essence identical. If the noise term is insignificant, then the linear expression is satisfied by the normal equations:

$$XX^{\mathrm{T}}\Psi = XY^{\mathrm{T}}. \tag{4.3}$$

The matrix product XX^{T} is invertible if the rows of X are independent, and we can express Ψ in terms of X and Y only (Strang, 1988, p. 156):

$$\Psi = (XX^{\mathrm{T}})^{-1}XY^{\mathrm{T}}. \tag{4.4}$$

Equation (4.1) may, however, involve significant noise levels and a "true" estimate of Ψ is then

$$\Psi = (X\Omega^{-1}X^{\mathrm{T}})^{-1}X\Omega^{-1}Y^{\mathrm{T}}, \tag{4.5}$$

where $\Omega = \mathcal{E}(\zeta\zeta^{\mathrm{T}})$ is the error-covariance matrix. One problem is that we only have an estimate of Ω if ζ is known ($\hat{\Omega} = \zeta\zeta^{\mathrm{T}}$). Ω may also be noninvertible. We can get around these problems by excluding the noise term from the analysis, and only attempt to predict the signal in Y that is related to X, which we refer to as $\hat{Y}$:

$$\hat{Y}^{\mathrm{T}} = X^{\mathrm{T}}\Psi. \tag{4.6}$$

By applying PCA to the data and truncating to the kth leading EOF, we also remove noise in X and ensure that $(X^{\mathrm{T}}X)^{-1}$ is invertible by writing the matrix in terms of its PCA products ($X = E\Sigma V^{\mathrm{T}}$),

$$[XX^{\mathrm{T}}]^{-1} = [(E_{(k)}\Sigma_{(k)}V^{\mathrm{T}}_{(k)})(E_{(k)}\Sigma_{(k)}V^{\mathrm{T}}_{(k)})^{\mathrm{T}}]^{-1} = [E_{(k)}\Sigma^2_{(k)}E^{\mathrm{T}}_{(k)}]^{-1}.$$

Hence, Eq. (4.5) can be expressed as

$$\Psi = (E_{(k)}\Sigma^2_{(k)}E^{\mathrm{T}}_{(k)})^{-1}XY^{\mathrm{T}}, \tag{4.7}$$

In many cases, colinearity may be a problem for regression. Some packages, such as R, offer methods which are robust to colinearity (it is easy to test this by constructing the following example). By "colinear," we mean that the inputs are not independent, but several contain the same information (are correlated).

One solution to colinearity is to pre-process the data by PCA or EOF analysis and use the principal components as inputs. The regression coefficients are not changed by swapping the orders of the inputs:

```
> t <- seq(0,20*pi,length = 1000)
> n <- rnorm(1000)
> y <- sin(t)
> x1 <- 0.4*y + 0.01*n               #  (1/0.4 = 2.5)
> x2 <- 0.3*y + 0.02*rnorm(1000) #  (1/0.3 = 3.3)
> lm(y ~ x1 + x2)

Call:
lm(formula = y ~ x1 + x2)

Coefficients:
(Intercept)           x1           x2
  0.0002824    2.1528601    0.4600223

> lm(y ~ x2 + x1)

Call:
lm(formula = y ~ x2 + x1)

Coefficients:
(Intercept)           x2           x1
  0.0002824    0.4600223    2.1528601
```

4.4 Canonical Correlation Analysis

CCA is a statistical method for finding spatially coherent patterns in different data fields that have the largest possible temporal correlation (Wilks, 1995; Preisendorfer, 1988).

Canonical Correlation Analysis, or CCA, is one brand of linear multivariate methods, which can be employed to downscale one field given another (Chen and Chen, 2003). Thus, in this case, the predictand is no longer univariate, but a map of gridded data, mathematically can be denoted as $\vec{y} \rightarrow Y$. Thus, in the following, Y is our predictand, and is a matrix.

There are various ways to carry out CCA. Here, we will discuss two types: the "classical CCA" and "Barnett–Preisendorfer CCA." Both types aim to produce maps of spatial structure, which have the maximum correlation.

The climate data can be thought of as a linear superposition of spatially coherent patterns at any time, and the time evolution of each pattern is described by an index (the extension coefficients) that determine how much each pattern contributes to the climatic state. The CCA yields two sets of weights that give the combinations of the corresponding sets of patterns with the maximum temporal correlation.

In CCA, we attempt to find the spatial patterns that give the maximum temporal cross-correlation between Y and X.

4.4.1 *Classical CCA*

We express two data fields as

$$\begin{aligned} Y &= \mathcal{G}\mathcal{U}^{\mathrm{T}}, \\ X &= \mathcal{H}\mathcal{V}^{\mathrm{T}}, \end{aligned} \tag{4.8}$$

where $\mathcal{U}$ and $\mathcal{V}$ are known as the *Canonical variates*, and describe the time evolution that have the greatest possible correlations,[a] and $\mathcal{G}$ and $\mathcal{H}$ are the spatial patterns associated with these.

[a]The leading column of each Canonical variate holds the time series which have the highest possible correlation, and the subsequent columns *must* be orthogonal to the respective leading column (Bretherton *et al.*, 1992). The second Canonical variate would represent the highest possible correlation of the data if the first Canonical variates and corresponding patterns were excluded from the data. The third column gives the highest correlation if the first two leading Canonical variates were removed before analysis, and so on.

One important property of the canonical variates, $\mathcal{U}$ and $\mathcal{V}$, is that each canonical variate is uncorrelated with all the canonical variates in the opposite set with the exception of the corresponding canonical variate (Wilks, 1995, p. 400).

The rotation matrices L and R, also referred to as the canonical correlation weights or canonical correlation vectors, satisfy the following properties: $L^{\mathrm{T}}L = I$ and $R^{\mathrm{T}}R = I$ (Preisendorfer, 1988, p. 299).

Mathematically, the analysis can be posed as a maximization problem, which can be expressed in the form of an eigenvalue equation. The temporal correlations are given as

$$\mathcal{U}^{\mathrm{T}}\mathcal{V} = LMR^{\mathrm{T}} = C. \tag{4.9}$$

The correlation matrix M contains the correlation coefficients on its diagonal, and all off-diagonal elements are zero when the columns in U and V are optimally correlated. By using the fact that the transpose of the rotation matrices equals their inverse (Strang, 1988; Press *et al.*, 1989), Eq. (4.23) can be written as

$$\begin{aligned} CR &= LM, \\ C^{\mathrm{T}}L &= RM^{\mathrm{T}}. \end{aligned} \tag{4.10}$$

By operating C in the second part of Eq. (4.10), we get $CC^{\mathrm{T}}L = CRM^{\mathrm{T}} = LMR^{\mathrm{T}}RM^{\mathrm{T}} = LMM^{\mathrm{T}}$, and we can now rewrite the equations in the form of eigenequations where rotation vectors in the columns of L and R are the eigenvectors (Preisendorfer, 1988, p. 302):

$$\begin{aligned} (C^{\mathrm{T}}C)L &= L(MM^{\mathrm{T}}), \\ (CC^{\mathrm{T}})R &= R(M^{\mathrm{T}}M). \end{aligned} \tag{4.11}$$

In order to solve the eigenvalue equation, the normalized covariance matrices, which are subject to maximization, must be estimated:

$$\begin{aligned} C_{YY} &= YY^{\mathrm{T}}, \\ C_{XX} &= XX^{\mathrm{T}}, \\ C_{YX} &= YX^{\mathrm{T}}. \end{aligned} \tag{4.12}$$

The matrix product, C (a $n_r \times n_r$ matrix) is a normalized covariance matrix:

$$C = C_{YY}^{-0.5} C_{YX} C_{XX}^{-0.5}, \tag{4.13}$$

and can be diagonalized using the SVD algorithm (Press *et al.*, 1989; Strang, 1988):

$$C = LMR^{\mathrm{T}}. \tag{4.14}$$

In Eq. (4.14), L and R are left and right rotation matrices, respectively, which yield optimal weighted combinations of the original time series. M is a diagonal matrix with the canonical correlation values in descending order on its diagonal. The CCA maps, $\mathcal{H}$ and $\mathcal{G}$, can be calculated from the covariance and the rotation matrices:

$$\mathcal{H} = C_{YY}C_{YY}^{-0.5}L, \tag{4.15}$$

$$\mathcal{G} = C_{XX}C_{XX}^{-0.5}R. \tag{4.16}$$

CCA extension coefficients (describing time evolution) can be calculated from the rotation matrices and the original data:

$$\mathcal{U} = C_{YY}^{-0.5}LY, \tag{4.17}$$

$$\mathcal{V} = C_{XX}^{-0.5}RX. \tag{4.18}$$

4.4.1.1 *Linear relationships from the model based on CCA*

The statistical models described here are based on linear relationships between the predictors, and predictands, and can be expressed as (Heyen *et al.*, 1996)

$$\begin{aligned} \hat{Y} &= \mathcal{G}M\mathcal{V}^{\mathrm{T}}, \\ \hat{X} &= \mathcal{H}M\mathcal{U}^{\mathrm{T}}. \end{aligned} \tag{4.19}$$

The first part of Eqs. (4.19) can be written as

$$\hat{Y} = \Psi X, \tag{4.20}$$

where the matrix Ψ is the statistical model that can be used for prediction. The canonical variates, $\mathcal{U}$ and $\mathcal{V}$, the canonical correlation maps, $\mathcal{G}$ and $\mathcal{H}$, and the correlation matrix, M, form the basis for the statistical model.

The two data fields Y and X are related to these CCA products according to Eq. (4.8), and the canonical variate $\mathcal{V}$ can be estimated as $\mathcal{V}^{\mathrm{T}} = (\mathcal{H}^{\mathrm{T}}\mathcal{H})^{-1}\mathcal{H}^{\mathrm{T}}X$. Y can therefore be predicted from X according to

$$\hat{Y} = \mathcal{G}M(\mathcal{H}^{\mathrm{T}}\mathcal{H})^{-1}\mathcal{H}^{\mathrm{T}}X. \tag{4.21}$$

The projection of X onto Y gives the predicted values of Y, and is denoted by $\hat{Y}$. The CCA model is the matrix $\Psi = (\mathcal{G}M(\mathcal{H}^{\mathrm{T}}\mathcal{H})^{-1}\mathcal{H}^{\mathrm{T}})$, where $\mathcal{G}$ and $\mathcal{H}$ are the canonical patterns and M is the diagonal matrix with the canonical correlations along its diagonal.

In `clim.pact` there is a function for applying CCA to two data fields. We will use the term "classical CCA" in the meaning of CCA applied directly onto the fields themselves, whereas the "Barnett–Preisendorfer CCA" method refers to CCA applied to EOF results.

In `clim.pact`, the CCA analysis can be applied to both EOFs and field, and if the argument "`SVD = TRUE`" (default) is used, then the CCA is computed using SVD according to Bretherton *et al.* (1992) rather than the covariance matrix method (Wilks, 1995; Heyen *et al.*, 1996) reproduced in the box above.

In some cases, the matrices are computationally singular, inhibiting the calculations. In these cases, it is useful to apply a different algorithm bypassing the problems. Care should be taken, as it is not guaranteed that the different approaches yield the same results.

4.5 Singular Vectors

The singular vector decomposition (SVD) analysis is a method for finding coupled spatial patterns which have maximum temporal covariance, and employs a numerical algorithm which calculates left and right eigenvectors (Press *et al.*, 1989; Strang, 1988). It is important to stress that this numerical algorithm and the coupled pattern analysis described here are two different concepts, although both are referred to as SVD.

The coupled pattern analysis SVD method (hereafter, referred to as just SVD) is similar to CCA (Benestad, 1998a; Wilks, 1995; Bretherton *et al.*, 1992; Preisendorfer, 1988), but differs from CCA by the fact that the SVD finds spatial patterns with the maximum covariance, whereas CCA finds patterns with maximum correlation. In other words, the SVD models are less sensitive to weak signals with high correlation than the CCA models.

The notations employed here are the same as in Benestad (1998a), and the predictor and predictand data can be written as follows:

$$\begin{aligned} Y &= \mathsf{G}\mathsf{U}^{\mathrm{T}}, \\ X &= \mathsf{H}\mathsf{V}^{\mathrm{T}}, \end{aligned} \tag{4.22}$$

where U and V are referred to as extension coefficients and describe the temporal evolution of the spatial patterns described by G and H. The SVD time series, U and V, have similar variance as Y and X, respectively, and the singular vectors (spatial patterns) satisfy $\mathsf{G}^{\mathrm{T}}\mathsf{G} = \mathsf{H}^{\mathrm{T}}\mathsf{H} = I$.

The matrices U and V contain the time series of the SVD patterns along their columns so that the leading columns of the two matrices have the greatest possible covariance.

Mathematically, the SVD analysis can be posed as a maximization problem, which can be expressed in the form of an eigenvalue equation in a similar fashion as for CCA. The covariance matrix is calculated according to

$$XY^{\mathrm{T}} = C_{XY}. \tag{4.23}$$

The mathematical solution to the maximization problem is similar to that of CCA, but with C replaced by C_{XY}:

$$C_{XY} = L\mathsf{M}R^{\mathrm{T}}. \tag{4.24}$$

The rotation matrices, L and R, represent the actual spatial patterns that have maximum covariance. The expansion coefficients are given by the matrix products (Bretherton *et al.*, 1992):

$$\begin{aligned} \mathsf{U} &= LY, \\ \mathsf{V} &= RX. \end{aligned} \tag{4.25}$$

The linear relationship between the predictors and predictands employing SVD analysis products is given in Eq. (4.25):

$$\begin{aligned} \hat{Y} &= \mathsf{G}C_{YV}C_{VV}^{-1}\mathsf{V}^{\mathrm{T}}, \\ \hat{X} &= \mathsf{H}C_{XU}C_{UU}^{-1}\mathsf{U}^{\mathrm{T}}. \end{aligned} \tag{4.26}$$

In Eq. (4.26) we have included the scaling factors C_{VV}^{-1} and C_{UU}^{-1} so that $\hat{Y}$ accounts for as much variance as Y. If $X = Y$, then $\mathsf{G} \equiv \mathsf{H}$, $\mathsf{V} \equiv \mathsf{U}$, and $X = \mathsf{H}C_{XU}C_{UU}^{-1}\mathsf{U}^{\mathrm{T}} = \mathsf{H}\mathsf{U}^{\mathrm{T}}$, which implies[b] that $C_{XU} = C_{UU} = C_{XX}$.

The singular value decomposition extension coefficients can be expressed in terms of the spatial SVD patterns and the original data, $\mathsf{V}^{\mathrm{T}} = \mathsf{H}^{-1}X$, and the first equation in (4.26) can therefore be expressed as

$$\begin{aligned} \hat{Y} &= \mathsf{G}\mathsf{M}C_{XX}^{-1}\mathsf{V}^{\mathrm{T}}, \\ \hat{Y} &= \mathsf{G}\mathsf{M}C_{XX}^{-1}\mathsf{H}^{\mathrm{T}}X = \Psi X, \end{aligned} \tag{4.27}$$

[b] $X^{\mathrm{T}}\mathsf{U} = \mathsf{U}^{\mathrm{T}}\mathsf{U} \rightarrow X = \mathsf{U} \rightarrow X^{\mathrm{T}}X = \mathsf{U}^{\mathrm{T}}\mathsf{U}$.

where $\mathsf{M} = C_{XY}$. The expression for Ψ can be obtained from Eq. (4.27), where $\Psi = \mathsf{G}\mathsf{M}C_{XX}^{-1}\mathsf{H}^{\mathrm{T}}$. The optimal predictor combination can be found, as in Benestad (1998a), by a screening method, where only the EOFs that increased the cross-validation correlation scores are included in the optimal models.

4.6 Further Reading

Bergant and Kajfež-Bogataj (2005) proposed using a more advanced regression method for ESD, the so-called multi-way partial least squares regression or the "N-PLS" regression scheme. By using N-PLS regression, they reported slight but general improvement over ordinary regression-based ESD.

Abaurrea and Asín (2005) used logistic regression to model the daily rainfall as the occurrence model and generalized linear model (GLM) with Gamma error distribution as the quantity model. GLMs will be discussed further in Sec. 9.5.1.

Heyen *et al.* (1996) was one of the pioneers in using CCA for downscaling, and the same approach was used by Busuoic *et al.* (1999), Busuoic *et al.* (1999), and Benestad (1999a) for ESD-analysis, and more recently followed up by Bubuioc *et al.* (2006). Busuioc *et al.* (2001) used CCA to downscale precipitation for Sweden, but they found that the GCM (HadCM2) was not able to reproduce the complexity of all circulation mechanisms controlling the regional precipitation variability. However, the most important ones for winter and autumn were identified in the first CCA pair, and were nevertheless well reproduced by the GCM.

One example of ESD based on the SVD method is that by Oshima *et al.* (2002), who used it to downscale monthly mean temperature in Japan. Benestad (1998b) also applied the SVD method to downscale monthly temperature in Norway, but found that the skill of this exercise was comparable with that of CCA and MVR.

4.7 Examples

Simple ESD

```
> library(clim.pact)
> narp<-getnarp(ele=601)
```

```
> print(narp$name)
> th <- getnarp("Torshavn",ele=601)
> data(DNMI.slp)
> eof <- EOF(DNMI.slp,mon=1)
> ds<-DS(th,eof)
```

The lines above were used to produce Fig. 4.2.

MVR

In the example below, a multivariate regression is applied to two fields X (`DNMI.t2m`) and Y (`DNMI.slp`) according to $\hat{Y} = \Psi X$. The results (`mvr`) is a new SLP field predicted from the gridded two-meter temperature (Fig. 4.4).

```
> library(clim.pact)
> data(DNMI.t2m)
> data(DNMI.slp)
> eof.1 <- EOF(DNMI.t2m,mon=1)
> eof.2 <- EOF(DNMI.slp,mon=1)
> mvr <- MVR(eof.1,eof.2)                # faster and more accurate than
                                         # using fields as input
> mvr <- EOF2field(mvr)                  # invert EOFs into original field
> mapfield(mvr)                          # Plot a map (left)
> DNMI.jan <- catFields(DNMI.slp,mon=1)  # Extract January months
> plotField(DNMI.jan,lon=10,lat=60)      # Plot the time series (right)
> plotField(mvr,lon=10,lat=60,add=TRUE,col="red",lty=2)}
```

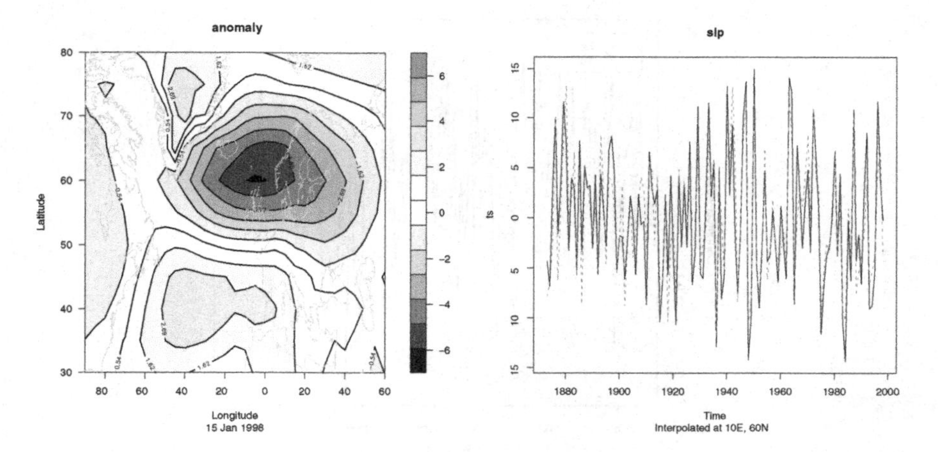

Fig. 4.4. Example of MVR results derived using the *R*-commands below.

4.7.1 *CCA*

Example 1: Simple CCA. CCA can easily be implemented in `clim.pact` (Figs. 4.5 and 4.6)

```
> library(clim.pact)
> data(DNMI.t2m)
> data(DNMI.slp)
> eof.1 <- EOF(DNMI.t2m,mon=1)
> eof.2 <- EOF(DNMI.slp,mon=1)
> cca <- CCA(eof.1,eof.2)
> summary(cca)
```

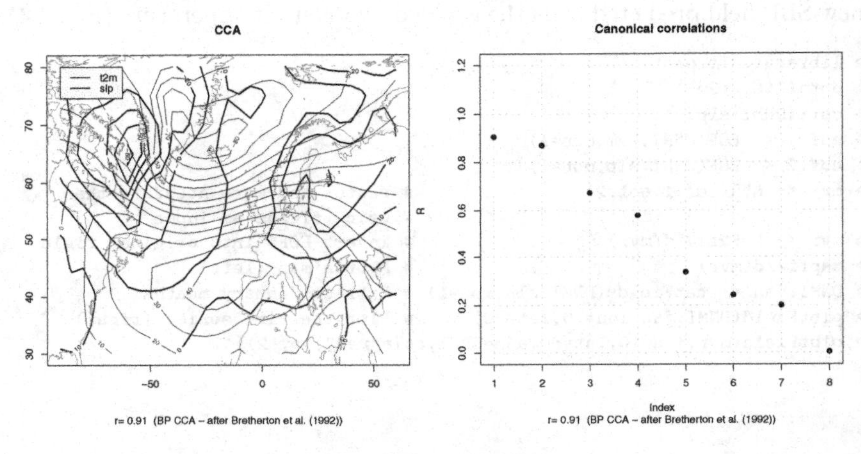

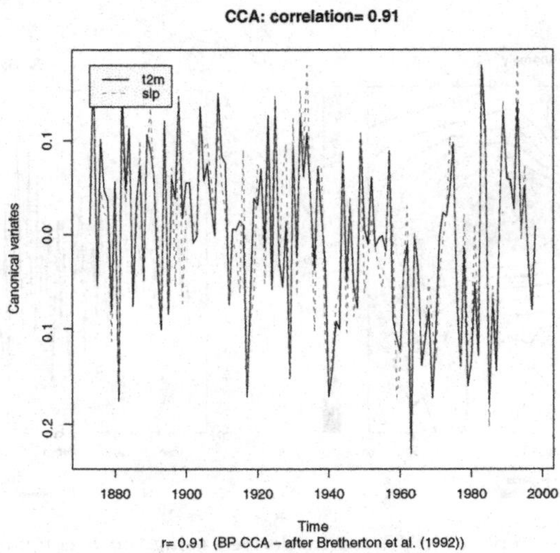

Fig. 4.5. Example of CCA results derived using the *R*-commands below.

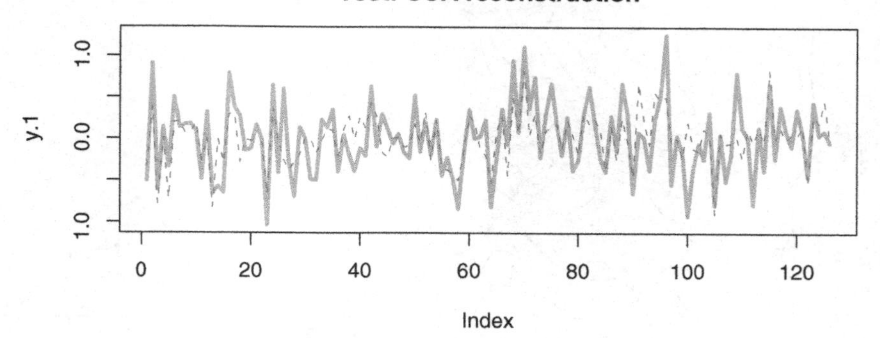

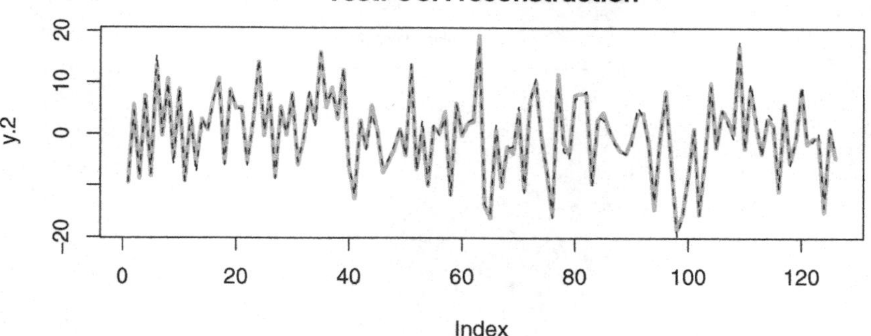

Fig. 4.6. Example of CCA test results.

Example 2: Simple CCA-based predictions (Fig. 4.7).

```
> library(clim.pact)
> data(DNMI.t2m)
> data(DNMI.slp)
> eof.1 <-  EOF(DNMI.t2m,mon=1)
> eof.2 <- EOF(DNMI.slp,mon=1)
> cca <- CCA(eof.1,eof.2)
> summary(cca)
> psi <-Psi(cca)
> ds.t2m <- predictCCA(psi,eof.2)
> DS.t2m <- EOF2field(ds.t2m)
> T2m <- EOF2field(eof.1)
> t2m.grd <- grd.box.ts(T2m,lon=10,lat=60)
> t2m.cca <- grd.box.ts(DS.t2m,lon=10,lat=60)
> plotStation(t2m.grd,mon=1,what="t",type="b",pch=21,lty=2,
            col="grey",trend=FALSE,std=FALSE)
```

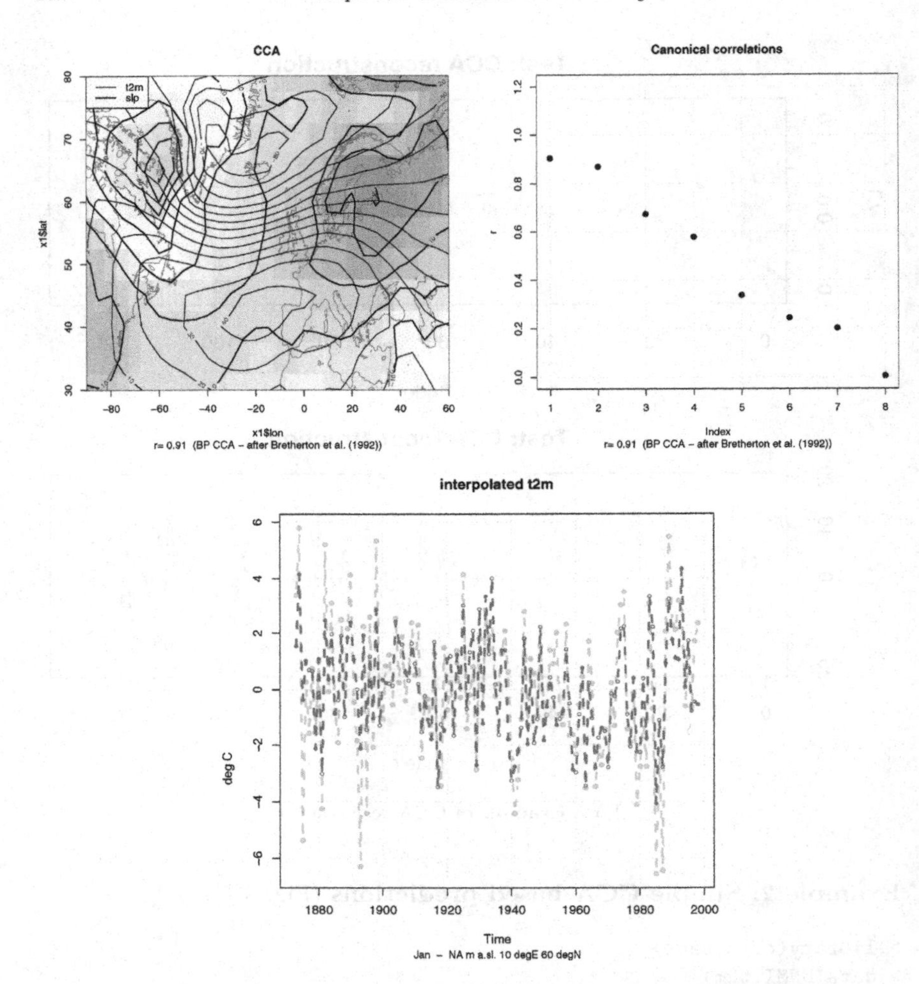

Fig. 4.7. Example of CCA results derived using the *R*-commands below.

```
> plotStation(t2m.cca,what="t",type="b",pch=21,lty=2,col="red",
              add=TRUE,trend=FALSE,std=FALSE)
> dev.copy2eps(file="cca-demo1.eps")
```

Example 3: CCA-based predictions with station data (Fig. 4.8)

```
> library(clim.pact)
> t2m <- stations2field()
```

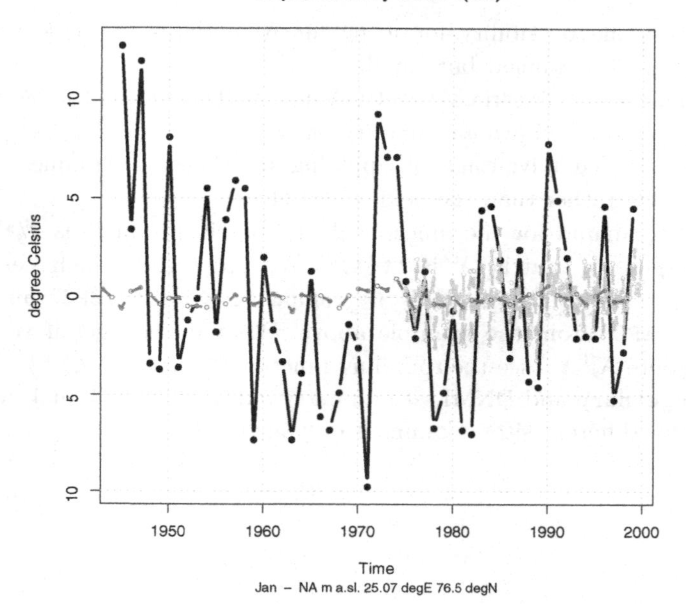

Fig. 4.8. Example of CCA results derived in a form of ESD using the *R*-commands below and taking station data as predictands.

```
> data(DNMI.slp)
> eof.1 <- EOF(t2m,mon=1)
> eof.2 <- EOF(DNMI.slp,mon=1)
> cca <- CCA(eof.1,eof.2)
> summary(cca)
> psi <-Psi(cca)
> ds.t2m <- predictCCA(Psi,eof.2)
> plotEOF(ds.t2m)
> DS.t2m <- EOF2field(ds.t2m)
> hopen.obs <- getnarp("Hopen")
> hopen.grd <- grd.box.ts(t2m,lon=hopen.obs$lon,lat=hopen.obs$lat)
> hopen.cca <- grd.box.ts(DS.t2m,lon=hopen.obs$lon,lat=hopen.obs$lat)
> plotStation(hopen.obs,what="t",type="b",pch=19,lty=1,trend=FALSE,
              std=FALSE)
> plotStation(hopen.grd,what="t",type="b",pch=21,lty=2,col="grey",
              add=TRUE,trend=FALSE,std=FALSE)
> plotStation(hopen.cca,what="t",type="b",pch=21,lty=2,col="red",
              add=TRUE,trend=FALSE,std=FALSE)
```

4.8 Exercises

1. Discuss the possibility for using linear methods in cases where the response is nonlinear but small.
2. Explain which criteria MVR, CCA, and SVD optimize the fit (what do they try to maximize or minimize?).
3. What is the advantage of applying the linear techniques to EOF products rather than the fields themselves?
4. The command for the inner-product (matrix product) is "`%*%`." The transpose of matrix X is "`t(X)`." Write a script which performs a matrix multiplication $X^{T}Y$ by taking "`Y <- eof.1$PC`" and "`X <- eof.2$PC`" from the example above. What do you get if you try to compute $X^{T}X$ (use the rounding function "`round(..,4)`.")
5. Use **`getnarp`** and DNMI.sst and carry out simple ESD-analysis (using `DS`) for different sites. Comment on your findings.

Chapter 5

NONLINEAR TECHNIQUES

5.1 Introduction

What is meant by "nonlinear techniques"? Usually, the term refers to the character of the response to changes in the influencing factors and the mathematical nature of the function describing this response. A linear response is one in which the response is the same for an *equal change* in the forcing factors regardless of their initial value. A nonlinear response is one where a similar change in the influences may cause a different response, depending on the initial state.

The temperature of water may be regarded as a nonlinear function of the heat supplied. For instance, it takes one calorie (about 4.184 J) of energy to heat 1 g of water at 1°C as long as the temperature is below boiling point. At the boiling point (100°C at the normal sea-level pressure), the temperature does not respond as before, as the extra energy is now used for the evaporation (or the so-called "phase transition").

5.2 The Analog Model

What is an "analog model"? The analog model basically consists of picking the date in the past when the situation most closely resembled the day for which the prediction is made. To reiterate, the method basically consists of resampling past data which coincide with the large-scale circulation regime that corresponds most closely with the given state of the atmosphere (Wilks, 1995, p. 198). In essence, the analog model is a search in a historical archive, describing all weather/climate events (predictors) in the past together with the local measurements of the quantity of interest (predictand).

Fig. 5.1. Clouds represent one of the most commonly observed phenomena that results from nonlinear processes.

The question then is: how to determine the past events that are most similar to the one in question? One way to do this is to employ a so-called *phase-space* (Gleick, 1987), which in essence is a coordinate system in two or more dimensions. Figure 5.2 shows a two-dimensional phase space for a case, where any hypothetical large-scale state (we can refer to these as "events") can be described entirely by two numbers, or two predictors.

Each event is shown as a symbol plotted in the coordinate system. The past event most similar to a predicted event is the one which is closest in the phase space. We let $\mathcal{D}$ be the smallest distance between the predicted event and the historical events (shown as arrows in Fig. 5.2).

5.2.1 *Caveats*

One concern regarding the analog approach is that it is incapable of predicting new records — magnitudes outside the range of the historical batch — since the predicted values are taken from the archives of past observations. It is likely that extreme events may become more frequent in the future (Huntingford *et al.*, 2003; Horton *et al.*, 2001; Palmer and Räisänen, 2002; Frich *et al.*, 2002; IPCC, 2002; Skaugen *et al.*, 2002a;

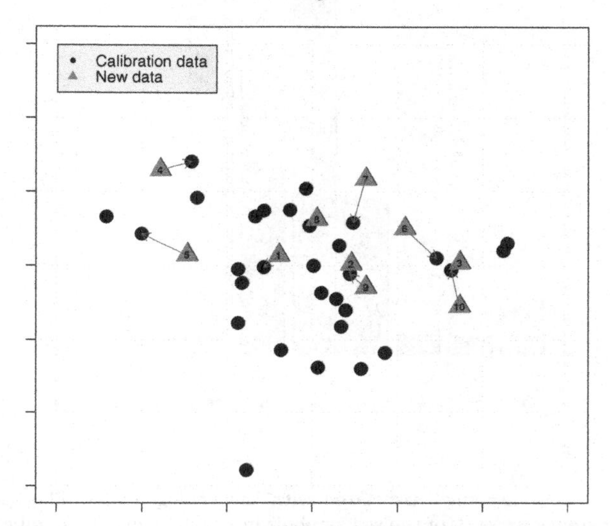

Fig. 5.2. Illustration of how the analog model works. In this example, the large-scale state can be described with two indices (variables). The black symbols represent events recorded in the past ("historical events"), whereas the red symbols represent predicted events.

DeGaetano and Allen, 2002; Prudhomme and Reed, 1999), and there is a non-zero probability of seeing new record-high values (Benestad, 2004d, 2003d).

The daily precipitation is not Gaussian, and the linear models are therefore not able to yield unbiased predictions of response to variations in the large-scale circulation. However, a least-squares approach may still, from a pure mathematical point of view, provide a solution to the regression coefficients that yields the lowest root-mean-squared-error, provided the coefficients are smooth functions with respect to the sums of the series (Press *et al.*, 1989, p. 555).

Linear models often involve a standard linear stepwise multiple regression (R function "lm"), whereas the analog model consists of a simple search for the nearest point in the principal components (PC) phase space. A linear model can be applied first to identify the PCs which are relevant, and then only the PCs retained in the stepwise search are used to define the PC phase space in which the analog search is performed. This also ensures that both linear and analog approaches use exactly the same predictors if they are to be compared (Fig. 5.3).

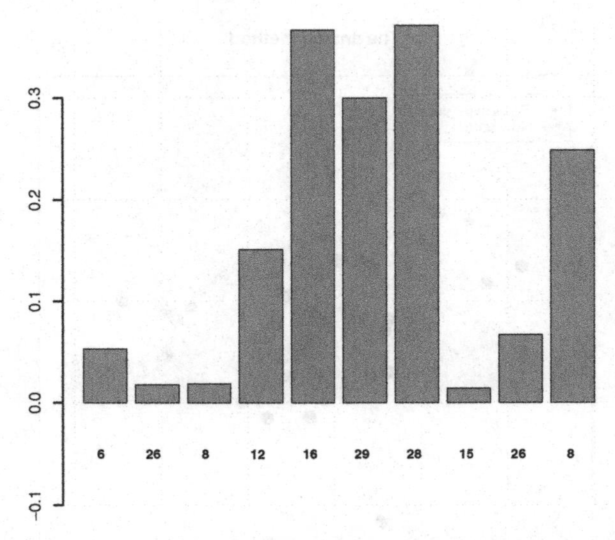

Fig. 5.3. The "book-keeping" associated with the analog model. The bar plot shows the smallest distance for each of the red symbols in Fig. 5.2, and the number underneath keeps track of which black symbol (day) was the closest.

If the analog model is to be used for studying extreme events for a future climate, it is necessary to modify the models so that they can extrapolate values outside the sample of observed magnitudes. Thus, the tails of the distributions present a problem if linear models suggest a more dramatic increase in the frequency of extreme warm temperatures than do the analog and combined methods (Imbert and Benestad, 2005).

In some cases a combined approach, where the mean trend is derived seperately through linear methods and then superimposed on the analog results, yields distribution shifts toward higher values than a purely analog model, and hence is capable of making extrapolations, i.e. producing values outside the range of values in the calibration sample. Although this solution can, in theory, predict changes in extreme values as a result of a trend, it cannot account for changes in extremes due to an altered variability in, e.g., the large-scale circulation patterns. Moreover, a study by Katz and Brown (1992) has suggested that the frequency of climatic extremes depends more strongly on changes in the variability rather than changes in the mean climatic state, and the results from Imbert and Benestad (2005) suggest that a linear model projects an increase in the frequency of very warm temperatures that is disproportional to the changes in the mean value. Hence, the addition of a mean trend to the analog method results may not suffice for the studies of future extreme values.

A different solution to improving the downscaling with analog models has been suggested by Hanssen–Bauer (private communications) who proposed to include warmer seasons in the calibration of the model. For instance, the present analog model for the winter season could extend the search for the nearest point in PC phase space representing the winter months to also include spring, summer, and autumn months to account for a warmer future climate. The limitation of this approach is, of course, that this approach would not be applicable for models for the summer season. This approach will be evaluated in future studies.

The analog model appears to be more appropriate for daily rainfall than for downscaling of temperature. Imbert (2003) analyzed the variance predicted for the daily precipitation with linear regression and an analog model, and found that while the linear method underestimated the variance, the analog model could describe ~100% of the variance.

Imbert and Benestad (2005) looked at the sensitivity of the predictions to different choices of the analog model design. One question is whether the various components should be weighted by the eigenvalue when searching for analogs in the data-space defined by the principal components from the EOF analysis. The result of the test suggested that the use of weighted principal components improved the results. Imbert and Benestad (2005) also proposed an adjustment scheme to improve the results, by adopting a common-EOF approach and shifting the mean and variance of the principal components so that the part describing the control period matches with those of the observations. Moreover, the PCs ought to be scaled by the principal values from the EOF analysis in order for the analog model to give good results. Such a weighting puts more emphasis on meaningful (leading) EOFs and reduces the effect of noise (high-order EOFs).

The analog model, as per definition, is incapable of making predictions of values outside the historical sample, and thus is unable to predict new record-breaking values (Imbert and Benestad, 2005). Even for a stationary process, new record-breaking values are expected to occur with time, albeit at successively longer intervals. Furthermore, it is not guaranteed that the analog approach is able to provide a realistic description of the upper tails of the PDFs. One can, however, apply a simple iid-test (Benestad, 2003d, 2004d) to check whether the data is independent and identically distributed (iid).

Another shortcoming associated with analog models is that they do not ensure a consistency in the order of consecutive days if weather regimes are not well-defined; however, this may to some degree depend on the evolution in the PC phase space and clustering of past states as defined by the predictors.

5.3 Classification Methods

Classification methods can be compared to a tree, where the large branches represent the clusters/classifications and the small twigs represent the individual data (e.g. solutions or weather conditions). The illustration in Fig. 5.4 provides a natural case where the snow is clustered on some of the main branches.

Weather types involve a classification of the synoptic (weather at a spatial scale of $\sim$ 1000 km) weather conditions. Traditionally, weather types have involved the Lamb weather types (for the United Kingdom) or the Grosswetterlagen (Germany), based on experience and subjective analysis.

The idea of classification methods is that one type of weather (e.g. a "cyclonic regime") tends to bring one type of precipitation or temperature patterns. For western Scandinavia, a strong NAO implies more pronounced and persistent westerly onshore winds that bring moist maritime air into

Fig. 5.4.

the interior. The mountain ranges force the air to ascend in order to pass, causing the air to cool and release much of its moisture (orographically forced precipitation).

With the advent of digital computers and more powerful data analysis resources, more objective classification methods have made an impasse. These range from simple analog models to cluster analysis, and neural nets.

5.3.1 *Cluster analysis**

Recurrent patterns are often observed in the climate. Such phenomena are often referred to as weather or climate regimes, and cluster analysis may be used for identifying these. The cluster analysis categorize the data into groups whose identities are not known in advance (e.g. Wilks, 1995, pp. 419–428). It is common to use the PCs from an EOF analysis as input to the cluster analysis (see Fig. 5.5).

Cluster analyses primarily represents an exploratory data tool, rather than an inferential tool. Commonly used cluster analysis procedures are hierarchical that build a set of groups, each of which is constructed by

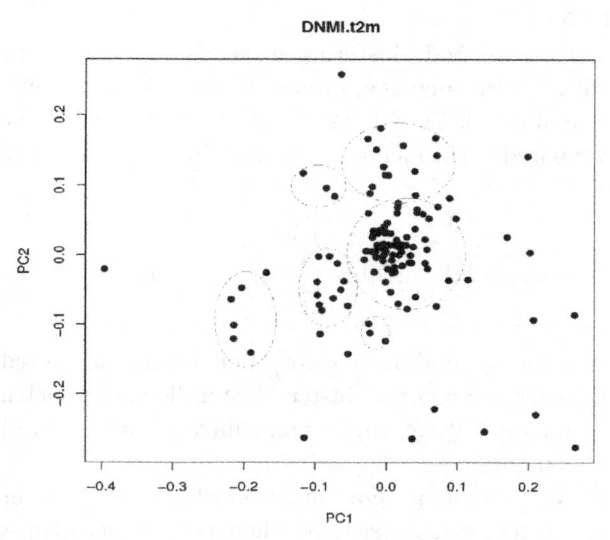

Fig. 5.5. Illustration of how clustering works. Scatter plot of two leading PCs of January SLP over the North Atlantic. The data can be organized into different classes according to their difference to the nearest cluster of points (here, marked by hand as vague red circles).

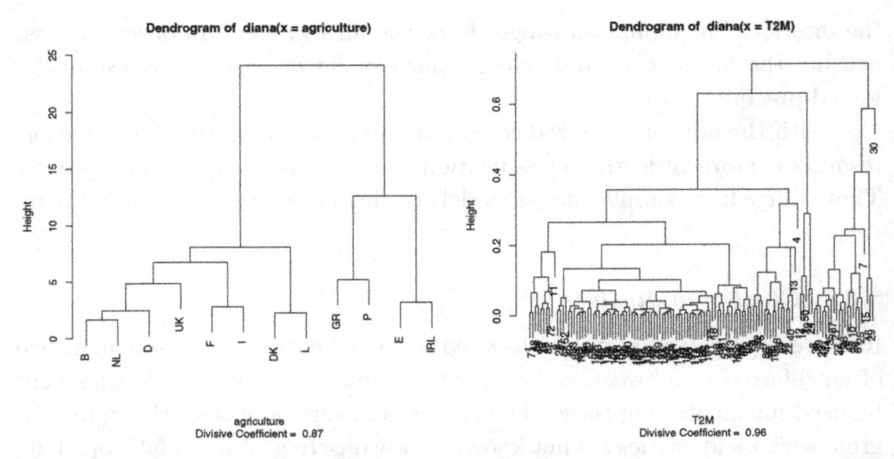

Fig. 5.6. Illustration of how clustering works. Left: an example from the `cluster`-package available on CRAN, "`example(diana)`." Right: a dendrogram for the two leading PCs of July T(2 m) over the North Atlantic shown in Fig. 5.5. The data can be organized into different classes according to different criteria (resemblance).

merging pairs of previously defined groups, starting with two groups. A *dendrogram* is a tree diagram showing the intermediate results from a cluster analysis (Fig. 5.6).

Distance measure and clustering methods may vary, but clustering bears similarities to the analog approach. Clusters tend to consist of points in a K- (e.g. number of EOFs) dimensional space, which are close to the each other compared to the members of other clusters. The distance between two clusters is

$$d_{j,i} = ||\vec{x}_i - \vec{x}_j|| = \left[\sum_{k=1}^{K}(x_{i,k} - x_{j,k})^2\right]^{0.5}.$$

Sometimes, the squared distance ($d_{j,i}^2$) or correlation is used. There are several methods for defining the cluster–cluster distances, such as (i) single-linkage, (ii) complete-linkage, (iii) average-linkage, (iv) centroid clustering, and (v) Ward's minimum.

One central question is: how many clusters are there in the data? Alternatively, when to stop the search? There is no universal answer to these questions, and subjective choice must be made, for instance, on summary statistics based on some kind of hypothesis testing or the inspection of the distances of the merged clusters.

The classification method is similar to the analog model in many respects, but a pool of historical data is now distributed to different classes according to the corresponding large-scale circulation pattern (Zorita and von Storch, 1997). Predictions can then be made by deciding which class a given large-scale situation belongs to, and then choose a random observation from the batch of data associated with this class. An analogy is to draw a number from one of several different hats, and the hat to be chosen is determined by the type of circulation.

5.4 Neural Nets

We will use the terms "neural nets" and "neural networks" here with the same meaning. Sometimes the term "artificial neural networks" (ANN) is used, meaning the same.

A neural network can be thought of as an algorithm that transforms an input vector $\vec{x}_i$ into an output vector $\vec{x}_o$ by stepwise nonlinear transformations. The number of steps, or levels, are often referred to as a layer of "neurons," and the intermediate layers are also known as "hidden layers." The number of layers can vary.

Neural nets and ANN can be employed in ESD group weather types into different classes (categories) as well as providing prediction models analogous to the linear models described in the preceding chapter. Nonlinear models offer the advantage of not being constrained to a linear relationship between the predictand and the predictor.

One disadvantage with neural nets is that they in general are complicated and do not offer a straightforward physical interpretation (Zorita and von Storch, 1999).

Neural nets also share a common drawback with other "deterministic models" in that they produce time series with less variance than in the observations, and therefore tend to underestimate the frequency or intensity of heavy rainfall and the frequency of dry days, even if the general distribution of dry and wet periods is reasonable (Fig. 5.7) (Zorita and von Storch, 1997).

5.5 Further Reading

A method for "correcting" (adjusting) the GCMs/RCMs was proposed and evaluated by Imbert and Benestad (2005). Then the question of how the

Fig. 5.7. One drawback with neural nets is the lack of transparency. They are often referred to as "black box" approaches.

results differ between the linear and analog models was discussed, and a combined-approach was proposed, where the trends derived from linear models are combined with the distributions from the analog models.

Zorita *et al.* (1995) proposed the "Classification and Regression Trees (CART) analysis," a type of classification method. However, when Zorita and von Storch (1999) compared the analog model with more complicated methods with a focus on the winter rainfall over the Iberian Peninsula, and found that the simple analog model was comparable in skill with methods such as CCA, a classification-based weather generator, and neural network.

Timbal *et al.* (2003) used analog models to predict daily extreme temperatures and rain occurrences. The models were able to partially reproduce observed climatic trends and inter-annual variability. Timbal *et al.* (2003) argued that the analog models are superior to direct model grid-average output.

Zorita and von Storch (1997) argued that when evaluating the analog method for a particular month, it is necessary to search the same season but for different years and use those for making predictions.

Sumner *et al.* (2003) also used an analog model to predict the daily rainfall for the Mediterranean Spain. They used a similarity index which utilized the Pearson product–moment correlation coefficients for the combined 925 and 500 hPa geopotential height fields.

Goodess and Palutikof (1998) used the Lamb weather type classification-based scheme to downscale daily rainfall in southeast Spain. Although the Lamb weather types were originally developed for the United Kingdom, Goodess and Palutikof argued that the scheme can also be successfully transferred to other regions.

Corte-Real *et al.* (1998) used a K-means clustering algorithm in conjunction with PCA in order to identify principal circulation patterns associated with daily precipitation over Portugal.

Zorita and von Storch (1997) argue that there has only been a limited number of neural network-based ESD studies, and that despite promising aspects, it remains to be demonstrated that they, in general, provide a useful downscaling method.

Penlap *et al.* (2004) used "self-organizing feature maps" (SOFMs) to group monthly precipitation from 33 stations in Cameroon into groups of stations with related temporal variability (homogeneous regions), followed by CCA to derive a statistical relationship between each of the groups and large-scale conditions. SOFMs can be considered as a type of neural nets, but in this case the method was used for selecting regions with similar characteristics prior to a linear model.

Schubert (1998) conducted a downscaling study with both linear as well as nonlinear statistical models, including nonlinear regression and neural networks. However, the nonlinear techniques did not improve the results significantly, suggesting that the relationships between synoptic circulation and local temperature, in this case over southeastern Australia, were mostly linear.

Crane and Hewitson (1998) used neural nets in ESD of daily subgrid-scale precipitation, and found that the downscaled precipitation increases were considerably larger than the change in the model's actual computed precipitation.

In a more recent study, Hewitson and Crane (2002) used self-organizing maps (SOMs) to identify the primary features in the synoptic scale SLP. They used SOMs to describe the synoptic changes over time and to relate the circulation to the local precipitation.

Schoof and Pryor (2001) also used ANNs and a cluster analysis to downscale daily maximum and minimum temperatures as well as total precipitation in Indiana, USA. The ESD involved rotated EOF as predictors and the downscaling was done using the PCs as input, cluster frequency in regression models and ANNs.

Unpublished work by Ramírez-Beltrán *et al.*,[1] has attempted to use neural nets to downscale atmospheric profiles based on radiosonde and satellite data.

5.6 Examples

5.6.1 *Analog model (Fig. 5.8)*

```
> library(anm)
> library(survival)
> library(clim.pact)
> data(oslo.dm)
> data(eof.dmc)
> anm.method <- "anm.weight"
> param <-"precip"
> a.lm <- DS(preds=eof.dmc,dat=oslo.dm,
>            plot=FALSE,lsave=FALSE,param=param,
>            ldetrnd=FALSE, rmac=FALSE)
> i.eofs <- as.numeric(substr(names(a.lm$step.wise$coefficients)
            [-1],2,3))
> a.djf <- DS(preds=eof.dmc,dat=oslo.dm,i.eofs=i.eofs,
>             method=anm.method,swsm="none",
>             predm="predictAnm",param=param,
>             plot=FALSE,lsave=FALSE,ldetrnd=FALSE, rmac=FALSE)
> plotDS(a.djf)
```

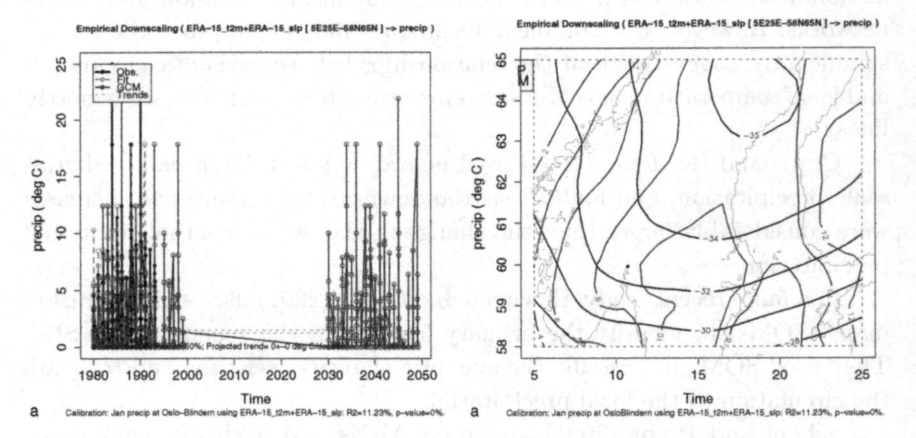

Fig. 5.8. Example of the results returned by the analog model.

---

[1] http://www.cmg.uprm.edu/reports/nazario.pdf

The example shows how the **anm**-package can be implemented. Here, a linear ESD is carried out prior to the analog model, in order to identify the EOFs likely to be important for the daily precipitation in Oslo. Then, the EOFs, not identified by the stepwise regression as important, are "discarded" before searching in the data space for analogs.

5.6.2 *Clustering*

```
> library(cluster)
> library(clim.pact)
> data(DNMI.t2m)
> eof <- EOF(DNMI.t2m,mon=7)
> plot(eof$PC[,1],eof$PC[,2],pch=19,xlab="PC1",ylab="PC2",
  main="DNMI.t2m")
> grid()
> T2M <- data.frame(x=eof$PC[,1],y=eof$PC[,2])
> plot(diana(T2M))
```

The example here was used to produce Figures 5.5 and the right panel in Fig. 5.6, and shows how clustering can be implemented for the climate data.

5.7 Exercises

1. Describe the analog model.
2. Take `DNMI.slp` and perform an EOF analysis for a chosen calendar month.
   (a) Generate a vector $\vec{x}$ of length similar to the number of PCs, but consisting of random numbers.
   (b) Use $\vec{x}$ to weight the EOFs and plot the spatial SLP pattern corresponding to the weighting.
   (c) Find the time t when the PC loadings are most similar to the weights in $\vec{x}$.
   (d) Find out the temperature in Oslo at time t (use `data(oslo.t2m)`).

   The exercises (a)–(d) are simple illustrations to how the analog model works.
3. Repeat the example above for Bergen, using weighted and nonweighted PCs.
4. Make a synthetic dataset consisting of random data (`rnorm()`) and make a `data.frame` object with the subvariables "x" and "y". Apply the clustering algorithm to these stochastic data. Discuss the results and compare with your expectations (what do you expect?).

Chapter 6

PREDICTIONS AND DIAGNOSTICS

6.1 Common Notations and Definitions

The **rejection level** α defines what is "sufficiently improbable," and is chosen prior to carrying out the tests. The level depends on the particular case, although 5% is a common threshold, and 1% or 10% are also often used.

The ***p*-value** is the specific probability that the observed value of the test statistics and all others which are at least as unfavorable will occur according to the null distribution.

The **rejection region**, or the critical region, is the tail (wing) of the PDF, which is outside the confidence limits.

One-sided and **two-sided** tests: The choice between the two depends on the nature of H_0. A one-sided test is used when there is an *a priori* reason to expect that either small or large test statistic (but not both) will violate H_0. An example is when the hypothesis is "It rains more in Bergen during September than in Oslo."

· A two-sided test is used when very large or small values of the test statistic are unfavorable to H_0. For instance, a two-sided test is used for H_0 = "The global mean temperature is influenced by the sunspots."[1] The null-hypothesis is then rejected if the test statistic p is greater than $100(1 - \alpha)/2$ (unit in %) or smaller than $100(\alpha)/2$ (in %).

Parametric tests and **nonparametric tests**: Parametric tests make some assumptions about the distribution of the data (usually theoretical distribution such as the Gaussian), and include Student's t-test and the likelihood ratio test. Nonparametric (distribution-free) tests do not assume

[1] Correlated or anticorrelated (high negative correlation coefficient).

Fig. 6.1. Layers in the rocks are not believed to form spontaneously, but are the results of changes in the environment. Armed with an understanding of physical laws, data, and statistical methods, it is then possible through "detective work" to make a statement about past events.

a theoretical distribution function, and include the rank tests such as the Wilcoxon–Mann–Whitney, resampling tests (bootstrap estimates), and Monte Carlo integrations.

6.2 Predictions

The predictions are usually considered as the "main" results; the actual number provides the answer to the question that we pose. Predictions are usually derived through regression or other linear methods, but as suggested above in the discussion of the diagnostics, they are not the sole results but are accompanied by the measures of the prediction quality.

The quality of the predictions depends on both how well the model describes the link between large scales as well as the GCMs ability to predict the predictor for the future. In this context, it is useful to reiterate the four criteria discussed in Sec. 2.2 of Chap. 2 that must be fulfilled in order to ensure reliable results:

(a) Strong relationship between the predictor and the predictand;
(b) Model representation;
(c) Description of change;
(d) Stationarity.

6.3 Trends

The downscaled results are often in the form of a time series. These can be subject to further analysis, such as testing for whether the results for one interval (e.g. the future) is significantly different from that of another (e.g. the past). Furthermore, a trend analysis can be used to examine the long-term behavior described by the predictions (e.g. a general warming or a cooling).

The simplest and most common way to study trends is to fit a linear trend, e.g. through an ordinary linear regression against time: $y(t) = y_0 + c_1 t$.

Benestad (2003e) argued that a linear trend may not necessarily be the best representation of the long-term evolution of a station series, and suggested fitting polynomials of the form $y(t) = y_0 + c_1 t + c_2 t^2 + c_3 t^3 + \cdots$ (Fig. 6.2) to give a better representation of the behavior when an increase or

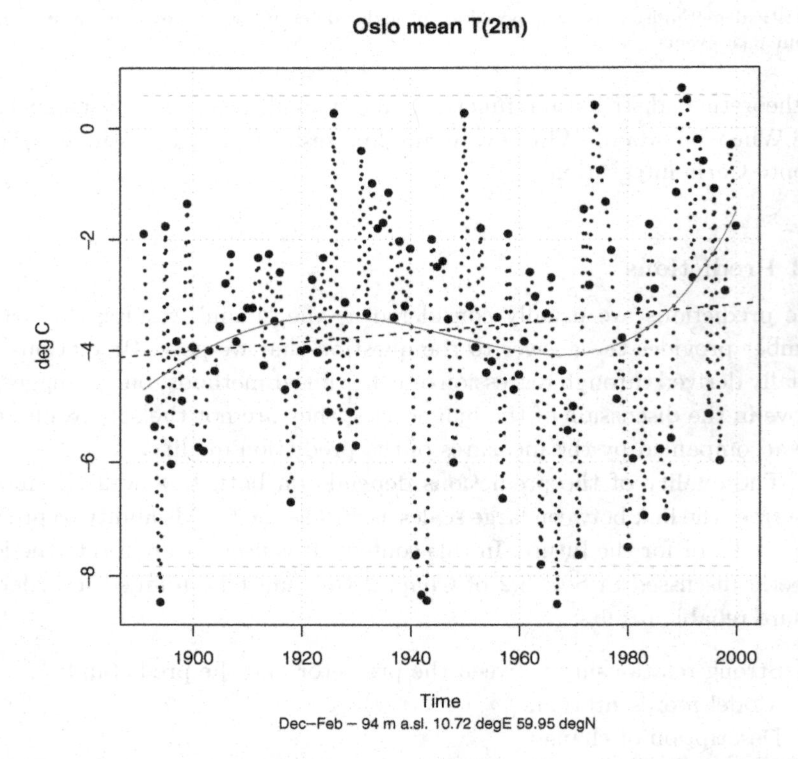

Fig. 6.2. A comparison between linear trend (blue) and a polynomial trend (red).

decrease has not been monotonic. The temperature evolution can resemble a cubic polynomial more than a linear trend, as seen in Fig. 6.2.

6.3.1 *Test for trends*

6.3.1.1 *Trend-testing based on Spearman rank coefficient*

Sneyers (1990) considered the rank correlation:

$$r_s = 1 - \frac{6\sum_{i=1}^{n}(y_t - i)^2}{n(n^2-1)}.$$

The null hypothesis was assumed to be that the distribution of r_s is asymptotically Gaussian with

$$E(r_s) = \mu = 0 \quad \text{and} \quad \text{var}(r_s) = \sigma^2 = 1/(n-1).$$

The null hypothesis is rejected for large values of $|r_s|$. Two-sided distribution, $\alpha_1 = P(|u| > |u(r_s)|)$, where $u(r_s) = r_s\sqrt{n-1}$, and H_0 is rejected at the level of α_0: $\alpha_1 < \alpha_0$. The trend-test is nonparametric, and more information can be found in Press *et al.* (1989), 13.8 (pp. 536–539).

6.3.1.2 *Mann–Kendall rank correlation statistics*

For each element x_i, the number of n_i elements x_j preceding it ($i > j$) is calculated so that $\text{rank}(x_i) > \text{rank}(x_j)$. The test statistic is calculated as $t = \Sigma_i n_i$. The distribution function of t is assumed to be asymptotically Gaussian with

$$E(r_s) = \mu = \frac{n(n-1)}{4} \quad \text{and} \quad \text{var}(t) = \sigma^2 = \frac{n(n-1)(2n+5)}{72}.$$

The Mann–Kendall test is two-sided, and H_0 is rejected for high values of $|u(t)|$:

$$u(t) = [t - E(t)]/\sqrt{\text{var}(t)}.$$

The trend-test is also nonparametric and suitable for identifying the interval in which the trend is most pronounced. See reference Sneyers (1990) and the section on Kendall τ in Press *et al.* (1989), 13.8 (pp. 539–543).

It is important to note that in some cases, the early observations were recorded with fewer decimal points than those at present. Often, the early temperatures were measured to the accuracy of 0.1°C, which for climate series can lead to many values with the same value: for instance many measurements of 12.1°C. It will be difficult to determine the exact rank of such series, as the measurements are not well resolved. A high number of equally ranked data points in the early part of the record may bias the trend analysis (Ø. Nordli, personal communications).

6.3.1.3 *Trend testing based on student's t-test*

The student's t-test may be used to test whether a linear trend is statistically significant. This may be done applying the one-sample test on the differenced series: $x_d(i) = x(i) - x(i-1)$ (time derivatives).

Alternatively, the t-test may be applied to estimate the linear slope from a regressional analysis: $t = (\hat{m} - m_0)/[\hat{\sigma_e}/(\Sigma_i(x_i - \bar{x})^2]$, where $\hat{\sigma_e^2} = \Sigma_i(y_i - \hat{y_i})^2/(n-2)$.

Often, error estimates and confidence intervals are provided by regression routines in numerical environments such as R.

6.4 Diagnostics

There are several diagnostics that can be obtained from ESD and may guide the interpretation of the quality of the results. First of all, there is the spatial distribution of weights, be it regression weights, CCA-pattern, or other types of loadings, depending on the chosen technique. The spatial structure of these weights should match the physical understanding of the situations affecting the local climate (Fig. 6.3).

For instance, for a case where SLP is used as a predictor for temperature or rain along the western coast of Norway, the weights should indicate a NAO-type pattern associated with the advection of mild and moist air from the west. Often, these weights should have a close resemblance to the maps of correlation coefficients (e.g. Fig. 1.2 in Chap. 1), which we will refer to as "correlation maps."

There are other statistics which also give an indication of how good the ESD-model is, such as the R^2 score, the F-statistic, and the p-value. These are standard statistics concepts and provide a measure of the closeness of fit or the theoretical likelihood that the series match by chance.

An additional diagnostic is the study of how the trends in the ESD-results vary with the seasons. Any abrupt changes may suggest that a

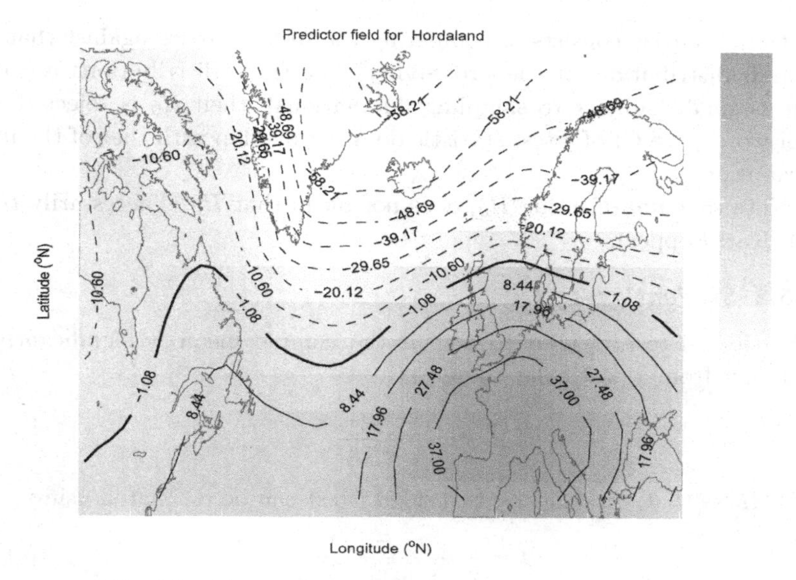

Fig. 6.3. The spatial pattern of weights for the predictor.

potential problem may be present. One reason may be that the chosen predictor domain is inappropriate — for instance, too large.

6.5 Statistical Inference

What is the likelihood that the results from an analytical test may be of pure coincidence? Is a high correlation between the sea surface temperatures in the North Sea and winter temperature in southern Norway really representative of the real world, or is it just a one-time fluke? A classical example of coincidental results is the high correlation between the number of storks and the birth rate in Germany (Höfer *et al.*, 2004). We will try to find ways to assess such cases and throw away such spurious results.

6.5.1 *Confidence intervals for rejection*

A confidence interval (CI) consists of a range of values that describe the sample uncertainty. The CI may sometimes be regarded as the "inverse" operation to hypothesis testing.

Typical use of confidence limits is the error bars shown for some sample statistics.

Data analysis consists of comparing the test statistics against that of the null distribution. If the probability that these distributions are the same is small (subject to sampling fluctuations), then H_0 is rejected. In other words, the CI of a test statistic does not overlap with that of the null distribution.

Note that *not* rejecting H_0, does not mean that H_0 is necessarily *true* (Ref. Karl Popper).

6.5.2 *Student's t-test*

One-sided t-test is used to test if the sample mean is significantly different from a particular value (μ_0):

$$t = \frac{\bar{x} - \mu_0}{\sqrt{s^2/n}}.$$

The H_0 is that "$\bar{x} = \mu_0$". A two-sided t-test can be estimated using

$$t = \frac{\bar{x}_1 - \bar{x}_2}{\sqrt{s_1^2/n_1 + s_2^2/n_2}}. \tag{6.1}$$

The t-test assumes normally distributed data for large numbers of data points, and are therefore inappropriate for daily precipitation which do not follow a Gaussian distribution.

For small numbers of data, the t-statistic follows a t-distribution:

$$f(t;N) = \frac{1}{\sqrt{N\pi}} \frac{\Gamma[(N+1)/2]}{\Gamma[N+1]} (1 + t^2/N)^{-(N+1)/2}.$$

The t-test formula assumes that each data point is independent of the other measurements, or that there is no serial correlation (autocorrelation) and that the two data sets are independent of each others.

If the values in the two data sets are correlated, then the denominator of Eq. (6.1) should be replaced by $\sqrt{s_1^2/n_1 + s_2^2/n_2 2\rho_{1,2} s_1 s_2/2}$, where $\rho_{1,2}$ is the Pearson correlation between x_1 and x_2. If the data are serially correlated (nonzero autocorrelation, $|a_1| > 0$), then the variance of each data set should be corrected using a variance inflation factor: $s_i^2/n_i \rightarrow s_i^2/n_i(1 + a_1)/(1 - a_1)$.

6.5.3 *F-statistics*

In regressional analysis, a relationship between two quantities, y_i and x_i, is sought, and there is a need for a measure of goodness-of-fit. Values of x_i may be used to make predictions for y_i, which are represented by symbol $\hat{y}(x_i)$. Linear models, $\hat{y}(x_i) = mx_i + c$, are commonly used in such types of analyses, but these methods also work for any model $\hat{y}(x_i) = f(x_i)$. The model errors, also called the *residuals*, are defined as $e_i = y_i - \hat{y}(x_i)$.

The *regression sum of squares* (sum of squared differences) is estimated according to

$$\mathrm{SSR} = \sum_{i=1}^{n} [\hat{y}(x_i) - \bar{y}]^2$$

and

$$s_e^2 = \frac{1}{(n-2)} \sum_{i=1}^{n} [y_i - \hat{y}(x_i)]^2.$$

The f-statistic, also known as the F-ratio, $F = \mathrm{MSR}/\mathrm{MSE} = (\mathrm{SSR}/1)/(s_e^2)$, and is a measure for the strength of the regression. Here, MSR is the predicted mean-squared-anomaly and MSE is the mean-squared-error.

A strong relationship between y_i and x_i gives a high F-ratio.

Other measures for goodness-of-fit is the coefficient of determination, R^2: $R^2 = \mathrm{SSR}/\mathrm{SST}$, where

$$\mathrm{SST} = \sum_{i=1}^{n} [y_i - \bar{y}]^2.$$

R^2 is 1 for a perfect regression and 0 for a completely useless fit (no correlation between y_i and x_i).

6.6 Quality Control

There are several ways of testing the quality. One important question is: how sensitive are the results to slight changes in the ESD setup? The answer to this question can be explored by varying the predictor domain or by varying the time interval.

We can also vary the predictor parameter to test the degree of sensitivity, but thorough physical understanding is still required in order to be able to interpret the results. It is expected that some predictors will produce different results than others, as discussed in Chap. 2, Sec. 2.2.

A useful approach is to examine the difference between the GCM results and ESD results. Large discrepancies may indicate inconsistencies. Furthermore, ESD results should be compared with the results from dynamical downscaling and vice versa.

When there is no *a priori* reason to believe that one method is better than the other for downscaling the results, it is expected that different approaches should provide a consistent picture. Diverging results prove that at least one of the methods introduce errors.

Downscaling methods and GCMs provide different levels of details, but substantial differences may suggest that there is a potential problem with some of the models.

Alternatively, there may be changes in the systematic structure of the climate system, such as changes in the snow-cover, sea-ice, cloud microphysics, lapse rate, vegetation, wave propagation, storm tracks, etc.

Thus, ESD in conjunction with other methods can be used to identify such (internal) changes as to how the various components of the climate relate to each other, and the issue of nonstationarity can be used to learn more about the system. In other words, ESD is an advanced level of data analysis.

Attempts have been made in the `clim.pact` package to automatically detect the size by examining the spatial correlation pattern. The function `objDS` has been designed for a complete treatment of the predictor choice and post-process quality control.

Post-process tests can be implemented for examining discontinuities in the seasonal cycle in trend estimates. We will refer to this kind of test as "seasonal trend test." Figure 6.4 shows one example based on the `objDS`-function (see example below). If the test identifies a discontinuity, the quality control routine recomputes the results for the months concerned, but with smaller predictor domain size in an iterative fashion.

Other quality controls involve testing the residuals for the time structure (autocorrelation, trends, etc.) of distribution. The residuals from a regression analysis should ideally have a Gaussian distribution consisting of iid data.

The `objDS`-function provides diagnostics showing the residuals, the predictor patterns, the R^2-statistics, p-value, and seasonal trend test.

For analog models (Chap. 10, Sec. 10.3), one diagnostic may be to plot the distance $\mathcal{D}$ with time.

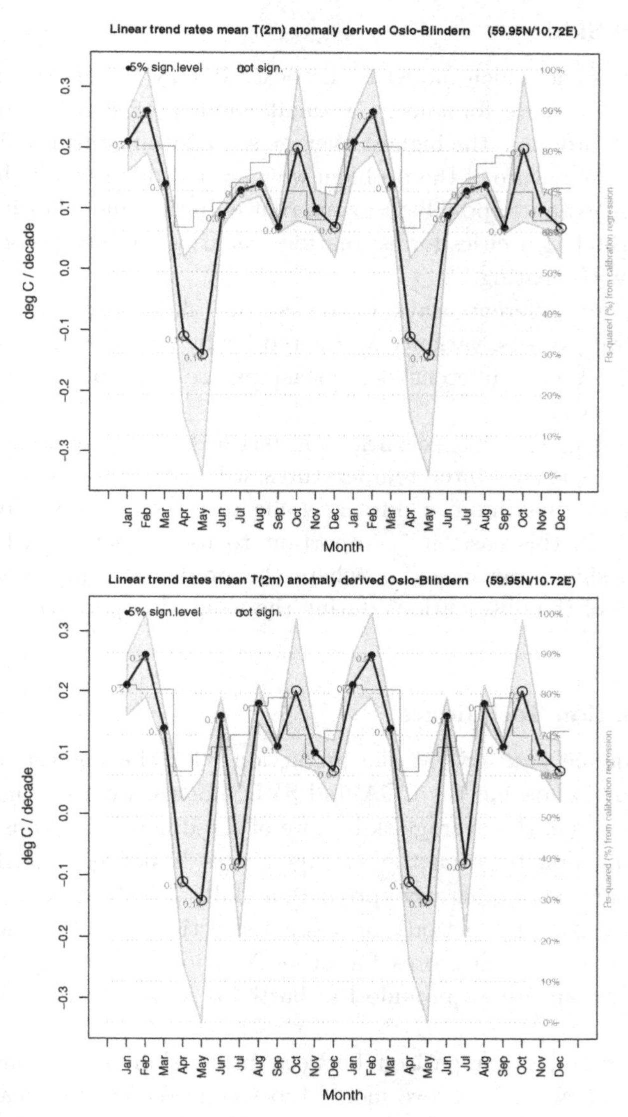

Fig. 6.4. Analysis showing the best-fit linear trend as a function of calender month (season), with trend estimate along the y-axis and the month on the x-axis. Here the year has been repeated, showing two cycles, in order to provide a good description of the change from December to January. The filled regions mark the confidence intervals of the trend coefficient estimates, and the blue curve shows the R^2 scores associated with ESD for the particular month. Top: results after quality control involving reiterations with smaller domains in the presence of jumps in trend coefficients; Bottom: before quality control.

6.7 Model Skill

How do we define which model is the best? It may be (i) the model that yields the best score for one or a small number of stations, or (ii) the model which produces the highest average score for all stations. The choice depends on the nature of the problem we want to address with the models.

It is, for instance, possible to construct empirical models with optimal skills near the larger cities for estimating scenarios for energy consumption associated with heating.

There are different ways of measuring skill, such as root-mean-squared (RMS) errors, variance accounted for by prediction, or correlation coefficients.[2] Again, different skill measures are appropriate for different types of forecasts.

In a global warming scenario, for instance, we may want to know how much the local winter temperatures will vary from year to year or how strong the maximum winds are going to be, e.g. the variance of the predictands. In this case, it is important to use models, which skillfully predict the signal variance (i.e. where the predicted signal accounts for about 100% of the observations during the validation period).

6.8 Evaluation Techniques

Often we define *best skill* as the prediction with the highest correlation score. When comparing the CCA and SVD models, we also employ mean station scores, i.e. the average skill score of the different stations.

A comparison between the average scores is not necessarily a good way of evaluating comparative prediction skill, as both model types may produce very good predictions for a small selection of stations and obtain mediocre- and low-skill scores for other locations. However, the station mean and variance may provide the basis for a crude test of the model differences.

Furthermore, a comparison between mean skill scores may give an indication of how well the two model types can predict large-scale climate anomalies.

[2]Other skill scores, such as linear error in probability space (LEPS) and the Brier Score for probability forecasts (Wilks, 1995) will not be discussed here.

6.8.1 *Anomalous correlation*

The anomalous correlation r is an ordinary correlation applied to the anomalies, rather than to the total value. The presence of a seasonal cycle in the data often swamps the year-to-year variations, which ESD tries to predict (the annual cycle is taken from the observations in `clim.pact` and then super-imposed onto the predicted anomalies).

A high r signifies a good match, or high degree of similarity, between two time series. The correlation measures the strength of the relationship between the predictor and the predictand, thus providing a description of how well the criteria [*a*] *Strong relationship between the predictor and predictand* in Chap. 2 Sec. 2.2 is met. Another way of visualizing the correlation is through a scatter plot, as shown in Fig. 6.5. The closer the points lie on the diagonal, the higher the correlation.

The correlation does not provide a measure of the magnitude of the differences, but only about the phase differences. Furthermore, r is insensitive to differences in the mean value.

CCA techniques in ESD aim to identify coupled patterns, which maximize the anomalous correlation between a set of station series and the predictors.

6.8.2 *The R^2 score*

Another measure of strength is the R^2 score, and the relationship between this score and the anomalous correlation is $R^2 = r^2$. The R^2 score provides a measure for the portion of the variance that ESD can reproduce. This is the standard measure of predicted variance used in ordinary regression analysis. The score is often expressed as a percentage, an R^2 score of 1 being 100%, which is a perfect reconstruction. If R^2 is 0, then ESD is unable to reproduce any of the signals in the predictand.

The proportional variance score is the ratio of the variance of the predictions to that of the observations, and is a measure of the predicted variance, but does not necessarily indicate how much *of the observed variability* can be described by the model. For instance, the predictions in some cases such as for analog models, the predictions may be associated with large variance despite being uncorrelated with the observations. In some circumstances, the variance score for a "perfect" prediction can be regarded as var = 100%.

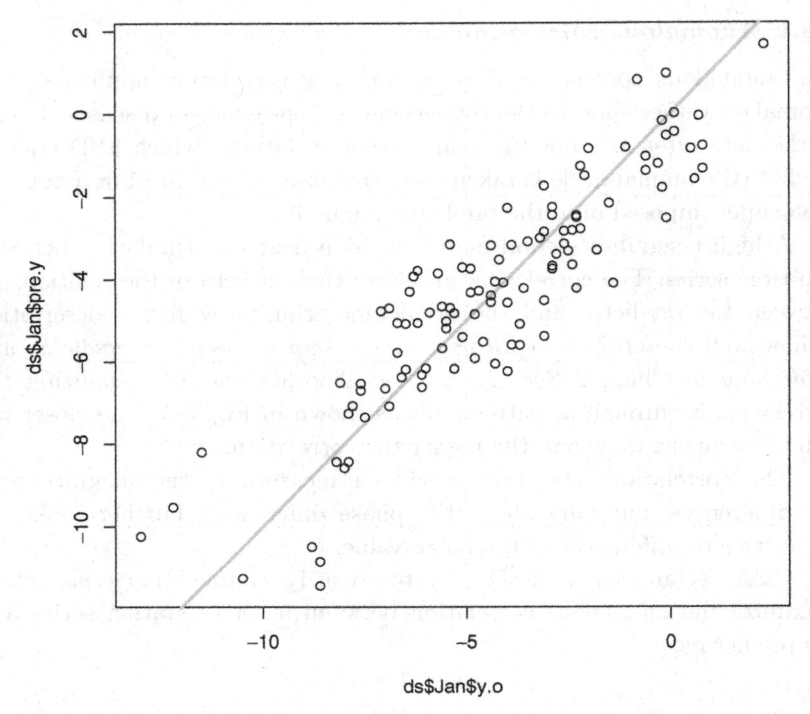

Fig. 6.5. An example of a scatter-plot, showing the empirical and corresponding ESD predicted values from the calibration period. A high correlation shows up as points close to the diagonal.

6.8.3 *Root-mean-square-error*

The root-mean-square-error (RMSE) is another measure of skill, and is computed from the difference between the predictions $\hat{y}$ and the measurements y according to:

$$\mathrm{RMSE} = \sqrt{\frac{1}{n}\sum_{i=1}^{n}(\hat{y}_i - y_i)^2}. \tag{6.2}$$

The larger the RMSE, the poorer is the forecast. The RMSE does not provide a clear indication about how much of the variance is recovered by the ESD or how well the statistical models manage to reproduce the variability. Furthermore, the RMSE is sensitive to both the differences in mean as well as differences in the extreme values (since RMSE is sensitive to the difference squared). Regression techniques aim to minimize the RMSE.

6.8.4 *Representative statistical distributions*

Standard measures, such as r, R^2, and RMSE, provide an index-based measure of the skill associated with different aspects of the predictions. There may be different requirements desired from an ESD analysis, depending on the situation.

Sometimes, it is the PDF that is important, rather than the phase information. None of the skill measures above provide a good measure of how well the PDF is recovered in an ESD analysis. Two variables with perfect correlation or R^2-score may, in principle, have different distributions (different means and standard deviations), and variables with nonnegligible RMSE may have the same PDF.

There are statistics that provide some description of the shape of the distribution, such as the skewness, kurtosis. In addition, a Kolmogorov–Smirnov test or χ^2-distribution may provide a means of assessing the similarity of the predicted and observed distributions.

6.9 Cross-Validation

The term *independent data* is used here to describe data which have not been used in model calibration (i.e. excluding the target years/months/days from the model calibration). Thus, the model does not "know" about these data when being constructed, and hence comparing the predictions against these gives a real indication of the model skill.

The cross-validation approach, also referred to as "jack-knife" in statistical literature, excludes one data point during the construction of a statistical model, and subsequently uses the model to predict the value of the predictand that was excluded from the model calibration (Wilks, 1995; Kaas *et al.*, 1998). The data not used for calibration of the model are referred to as *independent data.*

The process of excluding one data point is repeated N times, where N is the total number of observations, and for each iteration different data points are used as independent data. Figure 6.6 illustrates the process for the first two data points, where the red data points are the ones excluded during the model calibration and the blue symbols represent the data used to train the model.

The cross-validation method implies the construction of N different models, which are based on different combinations of the data. Cross-validation is also referred to as "jack knife" approach.

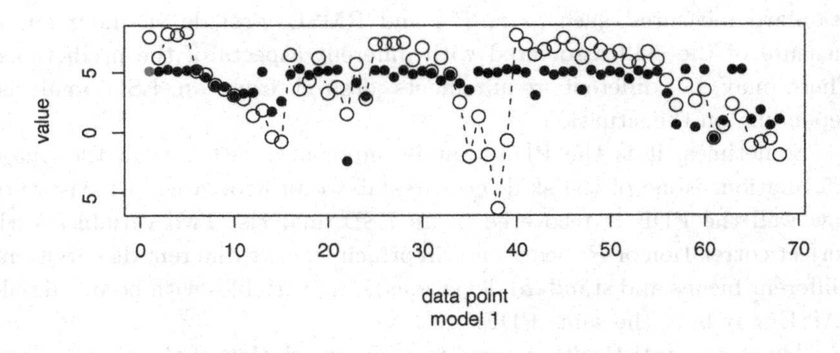

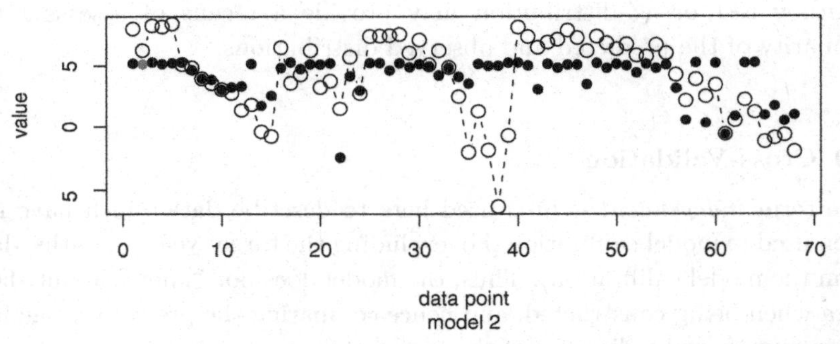

Fig. 6.6. Illustration of how cross-validation is implemented where the first data point is excluded from the model calibration (upper) and then the next point is excluded from the calibration batch (lower). The process continues until all the points have been excluded once, resulting in N different models and model predictions. The predictions for the independent data are subsequently combined to be evaluated against the corresponding model predictions. Blue symbols show the data used for model calibration, and red symbols represent data not involved in the training of the model.

Cross-validation may also be used to examine the data record. If the calibration provides a high R^2-score, but the difference between the predicted value and the empirical value at time t is great, then it may be an indication for errors in the observational record at time t.

There are several issues concerning cross-validation. For instance, which model should one use for making the final predictions? Should all the N data points then be used to make a "super-model" used for predictions? One could examine each of the coefficients from the N different models to see if there is a large scatter, and take the mean for each as the best estimate for the coefficients. Furthermore, the spread can be used for constructing error bars associated with the model parameters.

6.10 Further Reading

Osborn and Jones (2000), Hanssen-Bauer and Førland (2000), and van Oldenborgh (2006) used techniques associated with ESD to diagnose the causes of regional or local warming. Osborn and Jones (2000) fitted the local variable to the variations in the circulations, and then analyzed the residual series where the interannual to decadal "noise" has been reduced.

Hanssen-Bauer and Førland (2000) found that the most recent warming in Norway could be associated with the changes in the circulation, but the warming in the early 20th century could not be attributed to the changes in the airflow.

Benestad (2001a) also used a regression analysis to examine whether the warming in Norway could be attributed to a change in the NAO. One useful tool to assist the interpretation of ESD results involves statistical inference tests, discussed next.

6.11 Examples (Fig. 6.7)

6.11.1 *Illustration of cross-validation*

```
> library(clim.pact)
> data(addland)
> # Retrieve and pre-process the data:
> locs <- getnordklim()
> ns <- length(locs)
> mT <- rep(NA,ns); d <- mT; x <- mT; y <- mT; z <- mT
> for (i in 1:length(locs)) {
>    obs <- getnordklim(locs[i])
>    mT[i] <- mean(obs$val,na.rm=TRUE)
>    d[i] <- min(distAB(obs$lon,obs$lat,lon.cont,lat.cont)/1000)
>    z[i] <- obs$alt
>    xy <- COn0E65N(obs$lon,obs$lat)
>    x[i] <- xy$x; y[i] <-  xy$y
> }
> d[!is.finite(d)] <- 0
>
> # Demonstration of cross-validation:
> all <- data.frame(y=mT,x=x,y=y,z=z,d=d)
> lm.all <- lm(y ~ x + y + z + d,data=all)
> plot(1:ns,mT,cex=1.5,xlab="data point",ylab="value",main="Cross-validation",
     sub="model 1",type="b",lty=2)
> points(1:ns,predict(lm.all),pch=19,col="grey80",cex=0.9)
```

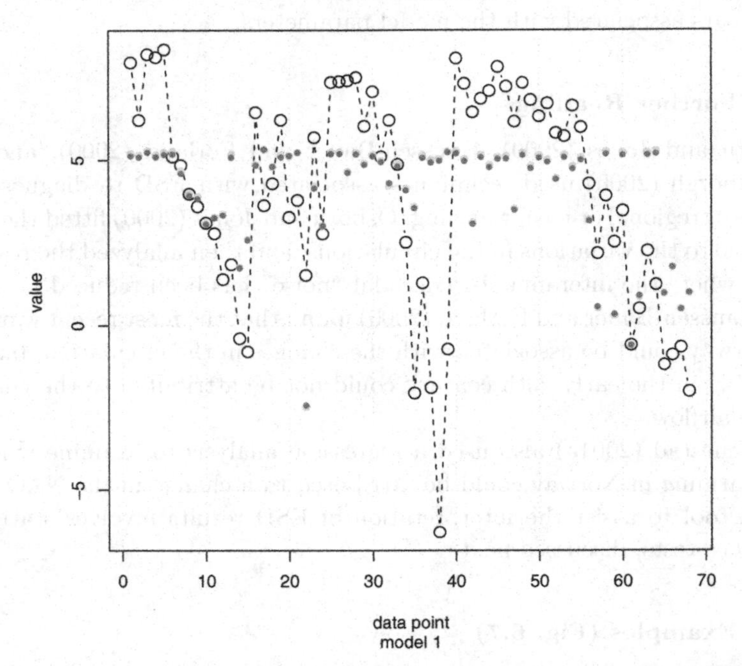

Fig. 6.7. Comparison: cross-validation (red) and best-fit based on the calibration sample (grey).

```
> grid()
> for (i in 1:ns) {
>   calibr <- data.frame(y=mT[-i],x=x[-i],y=y[-i],z=z[-i],d=d[-i])
>   x.pred <- data.frame(y=mT[i],x=x[i],y=y[i],z=z[i],d=d[i])
>   lm.xval <- lm(y ~ x + y + z + d,data=calibr)
>   points(i,predict(lm1,newdata=x.pred),pch=19,col="red",cex=0.7)
> }
>dev.copy2eps(file="cross-val-demo.eps")
```

6.11.2 *Using ESD to validate cyclone counts*

ESD has been used to examine the quality of the cyclone statistics in Benestad and Chen (2006). Since the number of low-pressure systems over a region scales with the mean SLP, one may expect to see an anti-correlation between the mean SLP and the cyclone number. Figure 6.8 shows that this was indeed the case.

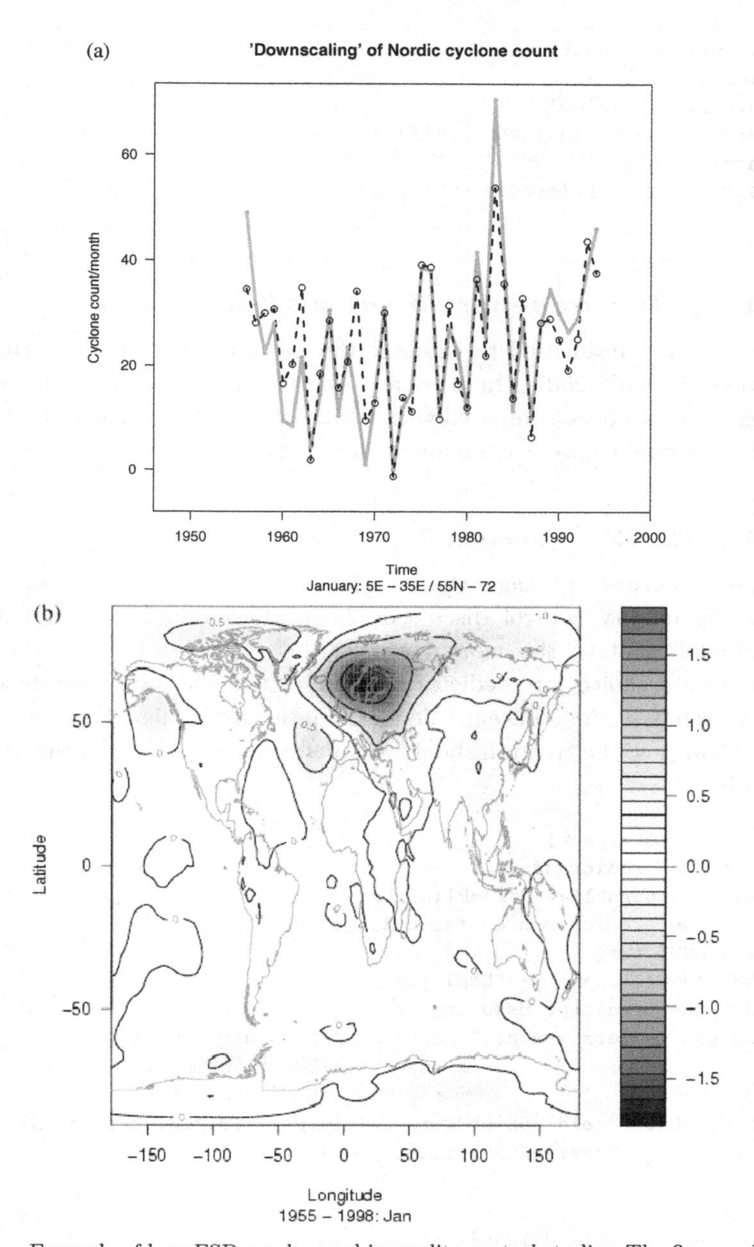

Fig. 6.8. Example of how ESD can be used in quality control studies. The figures show the ESD between the cyclone count over Fennoscandia and the gridded monthly mean SLP. Upper panel: predicted (black dashed) and observed (grey) counts; lower panel: predictor pattern. From Benestad and Chen (2006).

```
> library(clim.pact)
> library(cyclones)
> data(Storms.ERA40)
> nordic <- cyclstat(Storms.ERA40)
> data(eof.slp)
> ds.cyclone <- DS(preds=eof.slp,nordic)
```

6.11.3 *ESD to reconstruct historical climate records*

ESD has also been used to examine the precipitation trends in the past (Benestad *et al.*, 2007). In this case, the trend derived from the station series itself has been compared with the trend obtained when first applying an ESD-model to predict the local climatic parameter.

6.11.4 *The objDS-function*

In the clim.pact-package, the function objDS incorporates many of the ideas on quality control discussed above. This function is actually an "overhead" call to the more basic DS, but also searches for the most appropriate choice for predictor domain region and subsequently checks if the trends in the adjacent calender months are similar. Figures 6.9 and 6.10 show how the trend in the downscaled results varies with the calender month.

```
> library(clim.pact)
># Get the predictand:
> oslo<-getnordklim("Oslo-Blindern")
># Get the predictor used for calibration:
> data(DNMI.t2m)
> DNMI.t2m$filename <- "DNMI.t2m"
># Get the predictor used for projection:
> t2m.gcm <- retrieve.nc("~/data/mpi/mpi-gsdio_t2m.nc",
>                        x.rng=range(DNMI.t2m$lon),
>                        y.rng=range(DNMI.t2m$lat))
> ds <- objDS(field.obs=DNMI.t2m,field.gcm=t2m.gcm,station=oslo,
>             lsave=FALSE,silent=TRUE)
>
>
> ds.map <- plotDS(ds$Jan)
>
> map(ds.map,main="",sub="")
> # Save the graphics encapsulated graphics...
> dev.copy2eps(file="predictormap.eps")
```

(a)

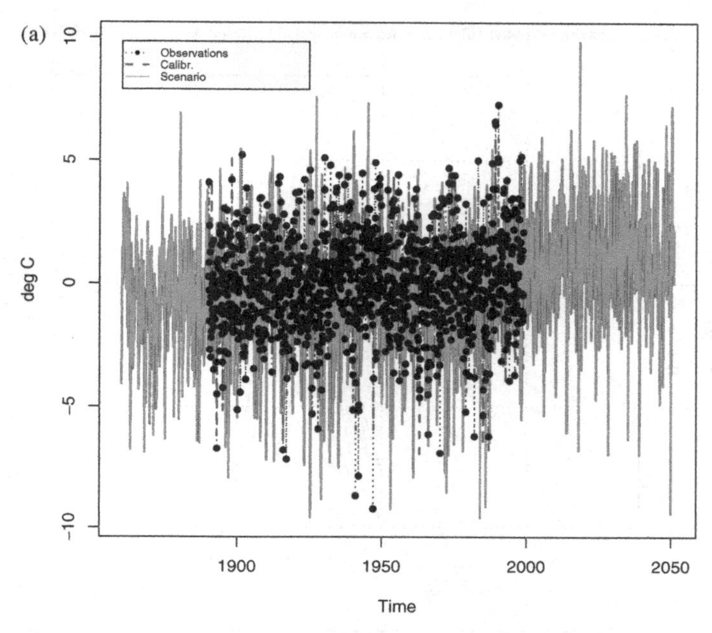

(b)

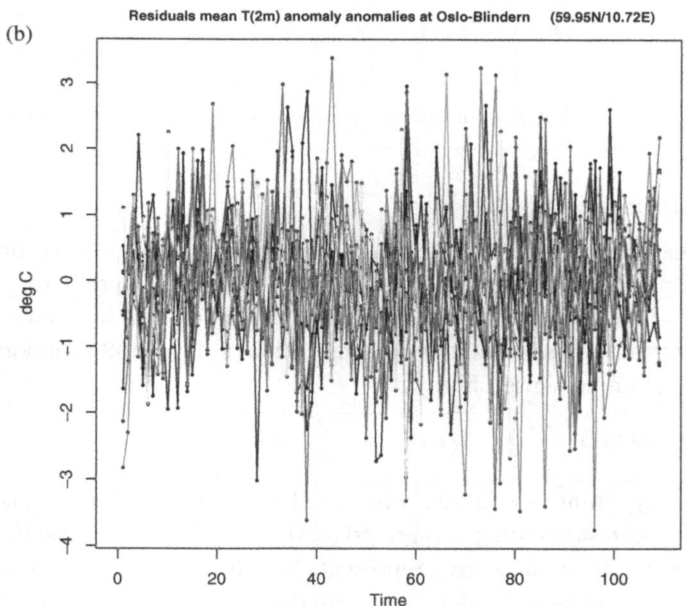

Fig. 6.9. (a) The main results from ESD, the downscaled time series (grey), shown together with observations (black); (b) the residual time series, one for each month.

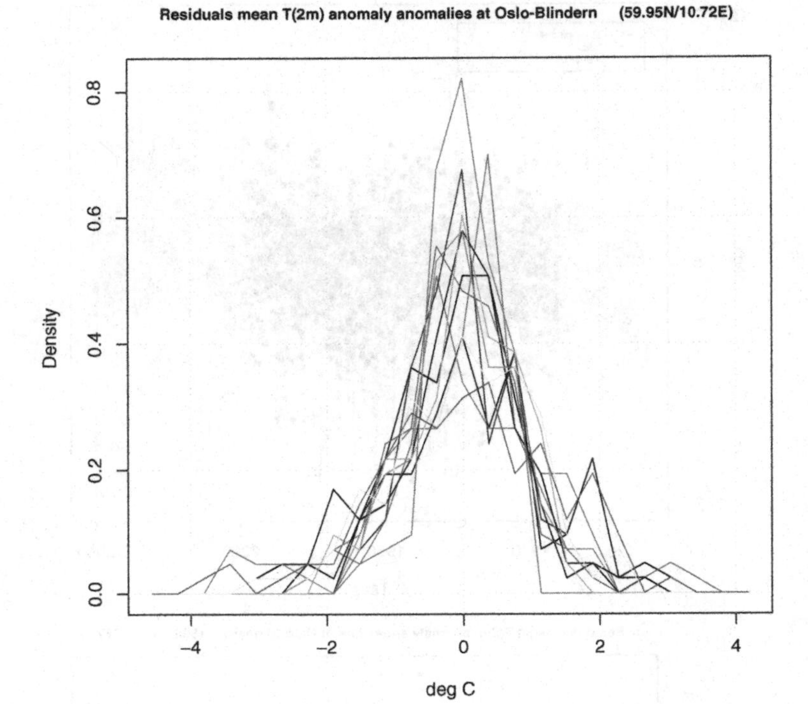

Fig. 6.10. The distribution of the residual data. Different colors are used for the different calendar months.

6.11.5 *Prediction pattern*

The `clim.pact` command `plotDS()` produced the diagnostics presented by the graphics in Fig. 6.11. Panel (a) shows the predictor pattern and (b) shows the downscaled series for the month of January. The following example assumes that demonstration of the `objDS`-function (above example) is implemented first.

```
> plotDS(ds$Jan)
```

It is important to consider whether the predictor pattern makes sense (here the greatest response is centered on the right location). The blue curve shows the empirical data, grey represents best-fit (i.e. dependent data), and the red curve presents independent predictions (here, based on GCMs). Also shown in red are the results from a trend analysis, both linear and polynomial.

(a)

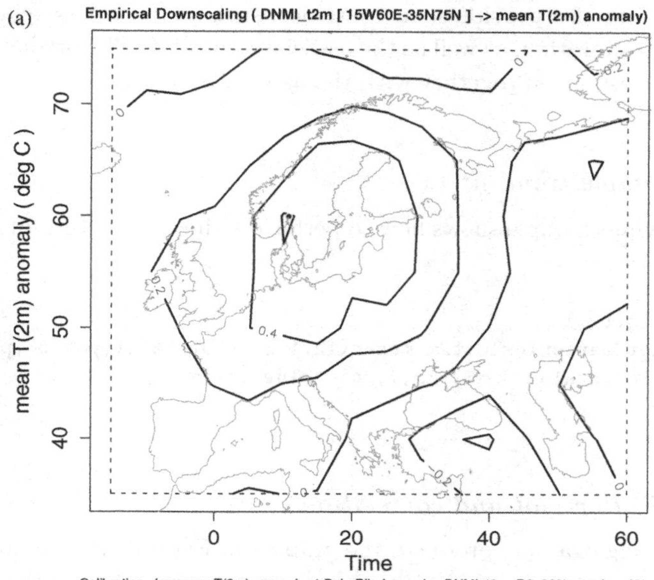

(b)

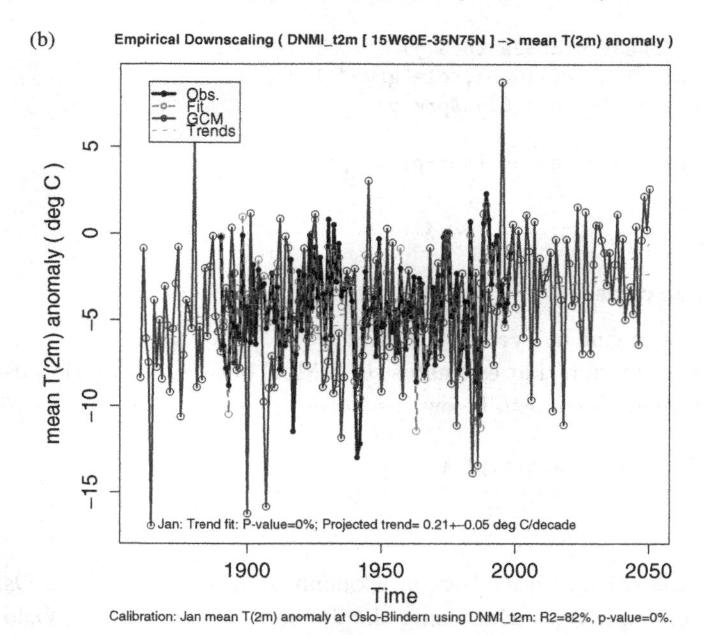

Fig. 6.11. Typical diagnostics produced by the call `plotDS`. The graphical results gives the p-value, the R^2 score and the trends, as well as visualizing the observations, the calibration fit, and the independent ESD-results.

The plots also provide some statistics, such as the R^2 associated with the model calibration, as well as the associated p-value. The predicted linear trends are also given together with the associated p-value.

6.11.6 *Simple trend-fit*

The following example shows how to perform a simple trend-fit for a station record:

```
> data(oslo.t2m)
> x <- plotStation(oslo.t2m,mon=c(12,1,2),what="t",type="b",pch=19)
> abline(lm(x$value ~ x$yymm,),col="blue",lty=2)
```

6.11.7 *Scatter plot and correlation*

The following example produces the graphic in Fig. 6.5. It is assumed here that the examples in Sec. 6.11 have been implemented before these lines.

```
> plot(ds$Jan$y.o,ds$Jan$pre.y)
> lines(c(-15,5),c(-15,5),col="grey",lwd=3)
> points(ds$Jan$y.o,ds$Jan$pre.y)
> grid()
> dev.copy2eps(file="scatter-plot.eps")
```

6.12 Exercises

1. What is meant by "cross-validation"?
2. Write a function that estimates the RMSE between two series, using the R-functions, as shown below.

   ```
   > RMSE <- function(x,y) {
   ...
   > }
   ```

3. Take the station series Bergen, Copenhagen, Stockholm, and Oslo, and do a multiple regression using the January temperature in Oslo as the predictand and the corresponding temperature for the other locations

as predictor. Use the function for RMSE (above) to evaluate the skill of the predictions.

4. Repeat the exercise above, but using a cross-validation approach. How do the results differ with the above?
5. Compute the EOFs of the DNMI SLP provided in the `clim.pact`-package for the January month. Take the PCs as input for a multiple regression against a station series. Repeat the above calculations with progressively smaller domains. Compare the results.

Chapter 7

SHORTCOMINGS AND LIMITATIONS

Shortcomings and limitations associated with ESD represent one source for uncertainties; however, it is important to keep in mind the fact that there is a range of sources associated with the projection of a future climate, as the uncertainty cascades from the estimation of future emissions/forcings, through biases and shortcomings associated with GCMs, to ESD and errors in the observations.

It is difficult, due to the complexity and convoluted nature of the models, to accumulate the errors through this chain of levels, and hence obtain a set of representative error bars at the end. One approach to estimate the errors is by running numerical experiments with different settings to estimate the effect of errors in the model parameters.

Although ESD introduces additional uncertainty, it is believed that it also has an added value by providing more realistic descriptions of the local statistics.

It is important to stress that the various downscaling approaches have different strengths and weaknesses and that one method cannot be universally considered as the "best." Skaugen *et al.* (2002b) have evaluated results for Norway from a nested RCM and they found that the RCM did not give sufficient realistic descriptions of the local climate as required by many impact studies. Empirical downscaling can, however, be used to provide more realistic local scenarios.

It is well known that linear regression (least squares methods) tends to yield lower variance than the original data (Klein *et al.*, 1959; von Storch, 1999). Figure 7.2 provides an illustration on how the range of values along the y-axis is reduced if the linear model (red line) is used together with the range of observed data values along the x-axis.

Fig. 7.1. Uncertainty are due to hidden details which may have a profound effect on the understanding of the problem. Here the mountains are hidden in mist.

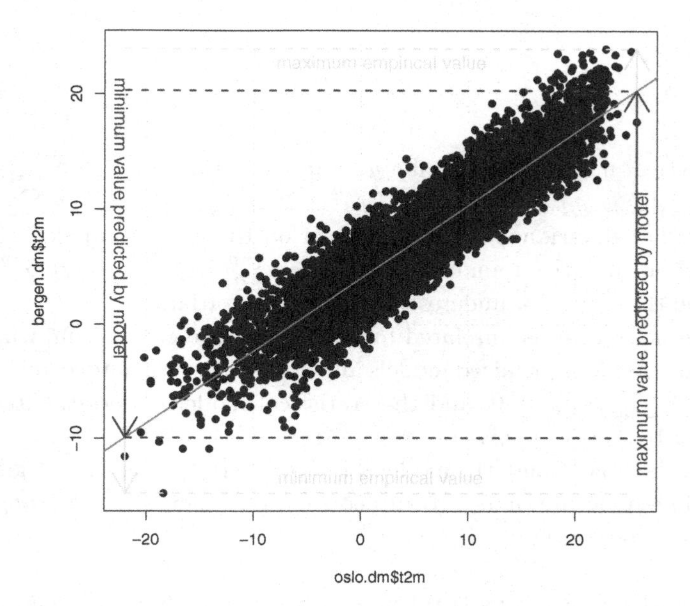

Fig. 7.2. Statistical models tend to underestimate the variance in the predictand.

One way to produce realistic variance levels in downscaling is to employ analog models (van den Dool, 1995; Zorita and von Storch, 1997, 1999; Dehn, 1999; Fernandez and Saenz, 2003) instead of linear regression.

In general, ESD suffers from two inherent limitations. The first is associated with the fact that empirical models that are approximate representations of links between large-scale and local-scale variations. Another limitation is associated with the assumption that the relationship identified with historical data would hold for the future. In the following, we discuss these limitations and point out ways to deal with the problems linked to the limitations.

7.1 Level of Variability

Zorita and von Storch (1997) argue that statistical models tend to only describe part of the behavior of the variable to be predicted. The part of the variability which is not reproduced by the statistical models is referred to as "noise" η, while the part that can be simulated is referred to as the "signal" y'. Thus, the variable can be regarded as a sum of the signal and the noise:

$$\vec{y} = \vec{y'} + \vec{\eta} \tag{7.1}$$

A statistical model will describe y', but the variance is $1/n \sum_{i=1}^{n} y_i^2 = 1/n \sum_{i=1}^{n} [y_i' + \eta_i]^2 = 1/n \sum_{i=1}^{n} (y_i')^2 + 2/n \sum_{i=1}^{n} y_i \eta_i + 1/n \sum_{i=1}^{n} \eta_i^2$. The predicted variance is the first term on the right-hand side of this expression, and the remaining terms $2/n \sum_{i=1}^{n} y_i \eta_i$ and $1/n \sum_{i=1}^{n} \eta_i^2$ describe the degree of underestimation of the variance.

If the noise term is unrelated to the signal (zero correlation, which is often the case for optimized models and noise with zero autocorrelation), then $2/n \sum_{i=1}^{n} y_i \eta_i \approx 0$, and the statistical model underestimates the variance by $1/n \sum_{i=1}^{n} \eta_i^2$.

If, on the other hand, the model is not properly optimized, or the noise has a time structure (non-zero autocorrelation), then $2/n \sum_{i=1}^{n} y_i \eta_i \neq 0$, and is non-negligible.

Since the climatology is taken as a baseline with a well-defined cycle, it is not needed to downscale this. In ESD, the downscaled anomalies do not describe the same magnitude as the empirical data, in other words, underestimating the variance. One fix to this problem has been the so-called inflation measures (von Storch, 1999), although this is generally not a good solution, as the part of the variations that linear statistical models do not capture cannot be associated with the given predictors. One source is commonly cited when justifying inflation (Klein *et al.*, 1959), but this paper says:

> "This [inflation] was done simply for the sake of obtaining a valid comparison between the two forecasting methods, and it should not be construed that this is necessarily the best objective method of forecasting temperature classes."

One reason may be that there may be unaccounted-for factors influencing the small-scale variable. When inflation measures have been invoked, the downscaled results have typically been scaled by the ratio of the standard deviations for the empirical data to the downscaled results.

There are some examples where the predictions have been "inflated" by scaling the variance of the predictions so that they describe the same variance as the original data. Zorita and von Storch (1997) argue that this is wrong because the results then are not consistent with the large-scale forcing.

von Storch (1999) also cautions against the use of "inflation" in ESD, however, it is also apparent from the simple maths done above that a simple scaling does not constitute a sound procedure, as this would entail neglecting the noise altogether through the process of scaling, yet saying the error is a problem — otherwise there would not be a need of rescaling.

Another way of looking at it is that we know for sure that the statistical model cannot account for the noise part, so if the prediction is rescaled, then we throw away this information.

The problem of reduced level of variance in the predictions is an important obstacle to predicting extreme events and exceedance over threshold values. Hence, other solutions are required to downscaling extremes, and these will be discussed in the later chapters.

7.2 Model Stationarity

We have already mentioned model stationarity, and by investigating the uncertainties in the model coefficient estimates from different calibration data combinations, a crude picture of how the models depend on the training period is emerging.

The fact that model skill varies with the season demonstrates that the relationship between predictors and predictands is not constant throughout the year, and the skill may also vary with the external conditions.

Another problem is that the PCA results for the present climate, on which our models are based, may not span the data space that corresponds to a global warming scenario. The issue of nonstationary EOFs may be a problem if there is a sudden change in the large-scale circulation in the future.

7.2.1 *"Perfect model" simulation*

So-called "perfect model simulations" (also referred to as "perfect model study") entail examining the relationship between the large and small scales in a GCM. One grid-point in the GCM is taken as the predictand and the region represented by several grid boxes as the predictor, and half of the simulations is used for calibrating the ESD-model describing the relationship between the large-scale GCM predictor and the grid-point predictand.

Any changes in the large-scale situation (predictor) and the local variability (predictand) simulated by the GCM, should, in principle, be identified in such a test. Figure 7.3 shows an example of such a test, and the divergence between the grid-point value plotted for the independent data and the ESD results, suggest that the SLP-based model is not stationary.

7.2.2 *"Historical" simulation*

In a "historical simulation" (Fig. 7.4) (also referred to as "historical study"), the relationship between the large and small scales in the past is examined. This type of exercise is analogous to the perfect model simulation, but now the gridded observations are split into calibration data and independent data for evaluation. Furthermore, a station series can be used instead of a grid-point value.

If the global temperatures for calibration period are lower than the validation period, and thus using a warmer validation period, this then

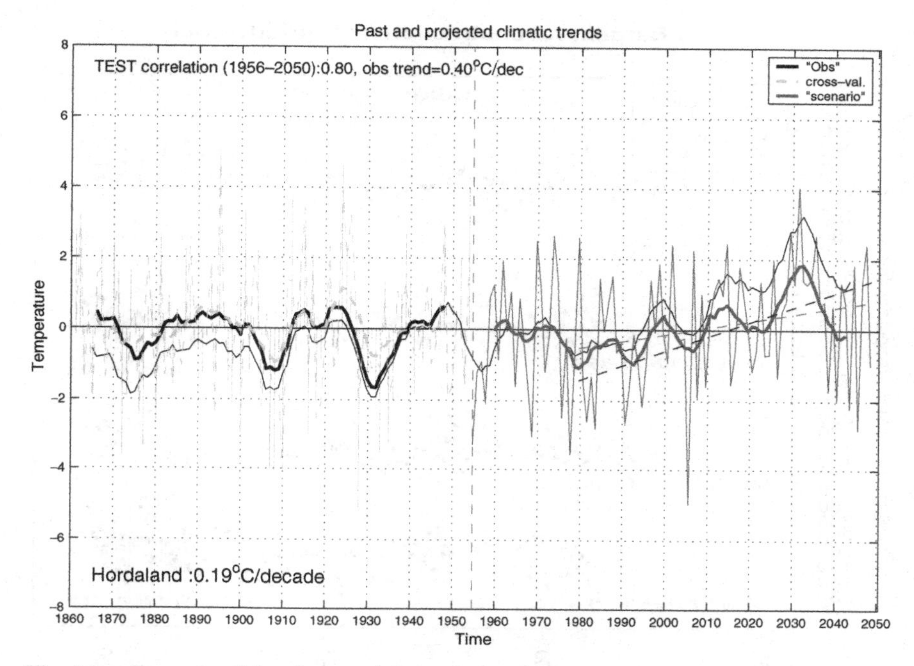

Fig. 7.3. Example of "perfect model simulations" where the large-scale SLP from the GCM has been used to (a) calibrate the ESD model for the first half of the simulation interval, and then (b) used as input to predict for the second half (Benestad, 2001b).

provides a good test to investigate the stationarity of the models in a slowly warming climate.

7.2.3 *Minimizing risk of nonstationarity*

If the ESD is based on physical considerations, taking all relevant factors into account, then one may in theory reduce the risk of nonstationarity. One should therefore always try to use predictors and predictands with the same physical units (see Sec. 2.2.2).

ESD should embody a physical causality, and ideally represent a physical expression:

$$y = x. \tag{7.2}$$

This could also be seen as a "dimensional analysis" common for the field of physics, where the object is to obtain an expression where physical units on the left-hand side of the equation matches those on the right-hand

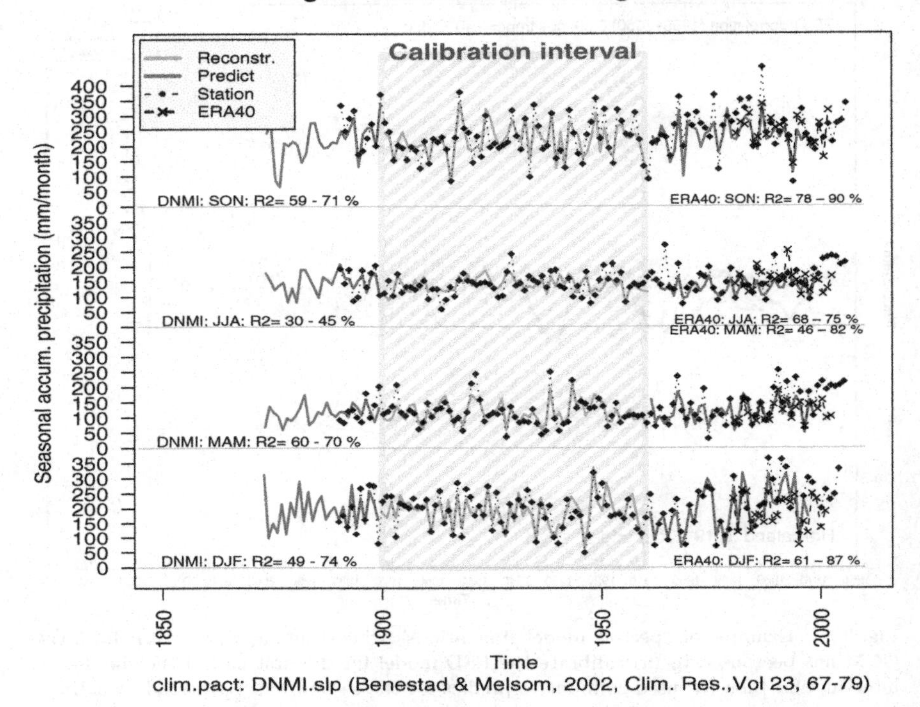

Fig. 7.4. An example of a historical simulation. The data in the interval shown as grey hatched region were used to calibrate the model and the data outside this region represent the independent evaluation data (Benestad *et al.*, 2007).

side. If the physical units are not the same, then the risk is greater that the relationship may be coincidental and not physics-based, and that the relationship may not hold in the future.

7.2.4 *Consistency*

One important aspect of ESD is the question whether the ESD results are broadly consistent with the GCM results. In principle, the ESD serves to refine the details, rather than producing a completely different picture. Thus, an appropriate consistency check may help answering the question if the ESD correctly represents the links in a different climate.

However, strong local effects and nonlinear relationship may exist, which can lead to considerable difference between ESD results and GCM simulations in some cases.

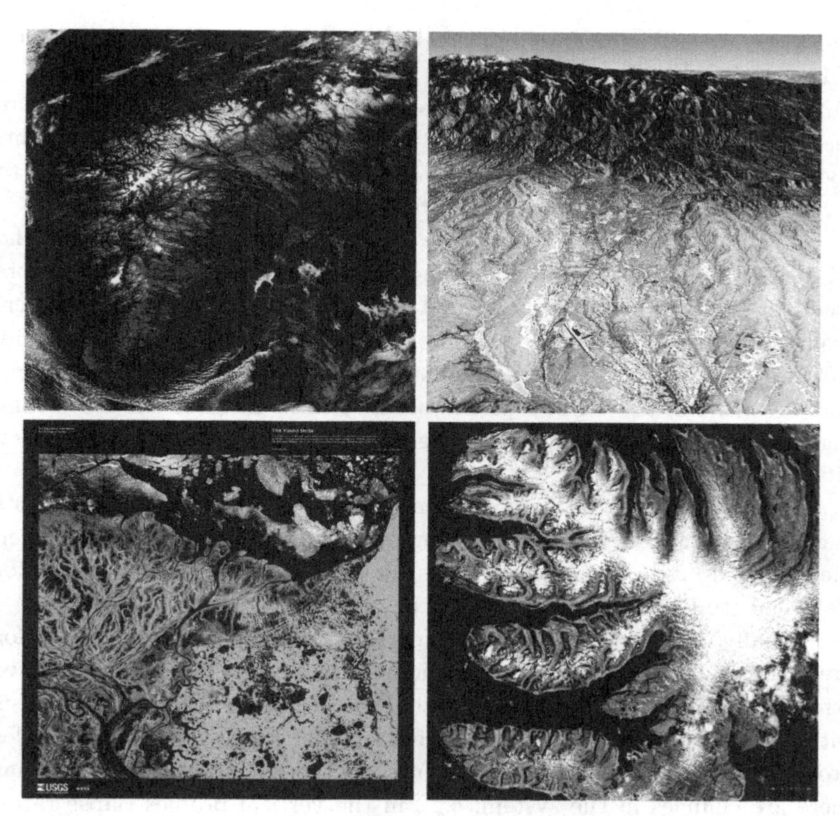

Fig. 7.5. Satellite images NASA/MODIS and Landsat (courtesy NASA) showing different landscape types. Different types may involve different local processes such as exchange of moisture, heat, momentum, aerosols, in addition to affecting the radiation budget (albedo).

The GCM's ability to represent the large-scale structures associated with the predictors is important. This consideration has led to the argument that the predictors should be taken from the free troposphere rather than the surface layer (Huth and Kyselý, 2000; Easterling, 1999), as the dynamic response (e.g., the geopotential heights) in the free troposphere is often perceived to be more skillfully reproduced by the GCMs.

Furthermore, climate models often have shortcomings when predicting the vertical profiles near the surface.

Surface processes furthermore tend to take place on small spatial scales, and are therefore often parameterized. Older climate models have treated surface processes with less sophistication than the state-of-the-art models, and there are still substantial uncertainties associated with the description of the land and surface influences.

On the other hand, ESD based on upper-air predictors may be more susceptible to nonstationarities if there are different trends at different heights (Benestad, 2005). Furthermore, the use of upper-air predictors may lead to very different results (trends) than surface predictors (Benestad, 1999c).

Diverging results also tend to be a sign of inconsistency between the GCM and the ESD results, and brings up the question of which is more credible. It may therefore be a good idea to apply ESD to both upper-level predictors and surface predictors just because these possibly may yield different answers. If the different results are inconsistent, then there may be errors in the GCM representation of these fields or the GCM may suggest that a climate change may alter old statistical relationships between different vertical levels (heights).

Thus, conducting ESD on different predictors and at different heights may shed more light on what is going on during a climate change. When the results do indicate inconsistencies, the exercise should be repeated for historical study as well as for perfect model study.

Usually, it is not possible to use corresponding physical quantities for predictors and predictand if the predictor variables are taken from the free atmosphere. Nevertheless, it is a useful test to repeat the ESD analysis with different predictor choices, e.g., from higher levels, if several variables provide a strong relationship. Differences in the results may indicate that there are changes in the system, e.g., in the vertical profiles (lapse rate), and hence provide further insight.

Another issue in ESD includes the choice of the region for which the predictor is representative (henceforth referred to as "predictor domain") and the question regarding the actual scale versus the size of the domain. The actual scale of the predictor does not change by choosing a larger predictor domain, and the spatial smoothness of the predictor variable should be an important aspect determining the latter.

7.2.5 *Take-home messages*

- ESD is based on the assumption of a strong relationship between the predictor and the predictand.
- ESD assumes that the statistical relationship between small-scale and large-scale parameters is stationary.

- ESD only brings added value when the large-scale parameters can be skillfully represented by the GCMs.
- The predictors *must* carry relevant signals, e.g., the global warming signals.

7.3 Examples

The code for making the plot, as given in Fig. 7.6 is given below:

```
> library(clim.pact)
> x.rng <- c(-10,40); y.rng <- c(50,75)
> load("ERA40_slp_mon.Rdata")
> load("ERA40_prec_mon.Rdata")
```

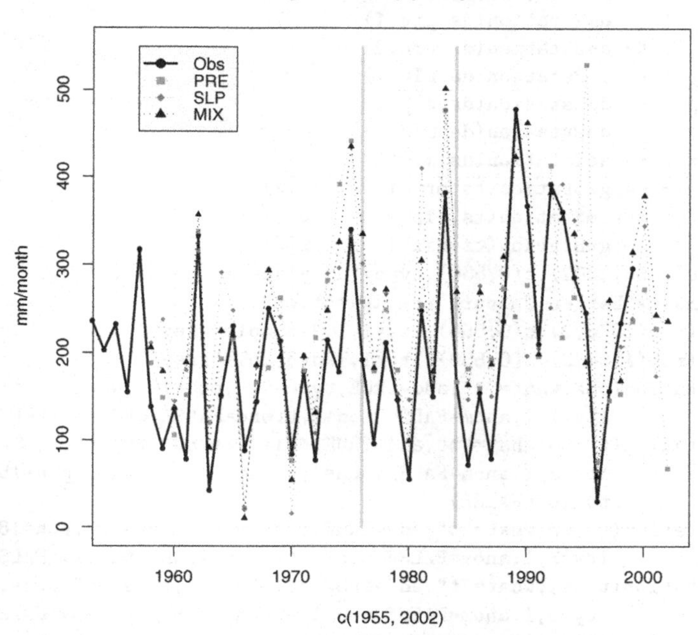

Fig. 7.6. Example of results obtained by splitting the data into two parts: one for calibration and one for independent evaluation. The analysis is done twice for each set of predictors, using swapping the calibration and evaluation parts, and then the predictions for the independent data are merged. The exercise is done for the large-scale precipitation as predictand, the SLP, and a mixed precipitation–SLP field.

```
> obs <- getnordklim("Bergen-Florida",ele=601)
> mix <- mixFields(prec,slp)
> eof.pre <- EOF(prec,lon=x.rng,lat=y.rng,mon=1,plot=FALSE)
> eof.slp <- EOF(slp,lon=x.rng,lat=y.rng,mon=1,plot=FALSE)
> eof.mix <- EOF(mix,lon=x.rng,lat=y.rng,mon=1,plot=FALSE)
> cal <- (eof.pre$yy < 1976)
> eof.pre$id.t[cal] <- "first"
> eof.slp$id.t[cal] <- "first"
> eof.mix$id.t[cal] <- "first"
> ds.pre.1 <- DS(obs,eof.pre,cal.id="first",plot=FALSE)
> ds.slp.1 <- DS(obs,eof.slp,cal.id="first",plot=FALSE)
> ds.mix.1 <- DS(obs,eof.mix,cal.id="first",plot=FALSE)
> cal <- (eof.pre$yy > 1984)
> eof.pre$id.t[cal] <- "second"
> eof.slp$id.t[cal] <- "second"
> eof.mix$id.t[cal] <- "second"
> ds.pre.2 <- DS(obs,eof.pre,cal.id="second",plot=FALSE)
> ds.slp.2 <- DS(obs,eof.slp,cal.id="second",plot=FALSE)
> ds.mix.2 <- DS(obs,eof.mix,cal.id="second",plot=FALSE)
> ts.pre.1 <- ds2station(ds.pre.1)
> ts.pre.2 <- ds2station(ds.pre.2)
> ts.slp.1 <- ds2station(ds.slp.1)
> ts.slp.2 <- ds2station(ds.slp.2)
> ts.mix.1 <- ds2station(ds.mix.1)
> ts.mix.2 <- ds2station(ds.mix.2)
> t.pre <- mergeStation(ts.pre.1,ts.pre.2)
> t.slp <- mergeStation(ts.slp.1,ts.slp.2)
> t.mix <- mergeStation(ts.mix.1,ts.mix.2)
> plot(c(1955,2002),c(0,550),type="n",ylab="mm/month",
>      main="Bergen January precipitation")
> lines(rep(1976,2),c(0,550),lty=1,lwd=3,col="grey")
> lines(rep(1984,2),c(0,550),lty=1,lwd=3,col="grey")
> plotStation(obs,what="t",add=TRUE,type="b",pch=16,lwd=2,
>             lty=1,l.anom=FALSE,mon=1,trend=FALSE,std.lev=FALSE)
> plotStation(t.pre,what="t",add=TRUE,col="grey50",type="b",
>             lty=3,l.anom=FALSE,mon=1,trend=FALSE,lwd=1,pch=15,
>            std.lev=FALSE)
> plotStation(t.slp,what="t",add=TRUE,col="red",type="b",pch=18,
>             lty=3,l.anom=FALSE,mon=1,trend=FALSE,std.lev=FALSE)
> plotStation(t.mix,what="t",add=TRUE,col="blue",type="b",pch=17,
>             lty=3,l.anom=FALSE,mon=1,trend=FALSE,std.lev=FALSE)
> legend(1957,550,c("Obs","PRE","SLP","MIX"),
>        col=c("black","grey50","red","blue"),lty=c(1,3,3,3),
>        pch=c(16,15,18,17),lwd=c(2,1,1,1),bg="grey95")
> dev.copy2eps(file="esd_ex6-1.eps")
```

7.4 Exercises

1. Discuss the various sources for uncertainty in ESD work.
2. Why is it a good idea to use more than one predictor variable?
3. Write a short R-script to downscale the January temperature in Oslo for the period 1980 and onwards, using only the data prior to 1980 for calibration. Use `data(oslo.t2m)` and `data(DNMI.slp)` to provide the predictand and the predictor. Plot the results. Use the example given above as a guide to writing the code.
4. Repeat the exercise above, but now use interpolated values taken from the gridded temperature as predictors instead of actual station measurements. Use the lines provided below as a guide to interpolate gridded T2m to the coordinates of Oslo.

```
> data(DNMI.t2m)
> data(oslo.t2m)
> oslo.dnmi <- grd.box.ts(DNMI.t2m,lon=oslo.t2m$lon,lat=oslo.t2m$lat)
```

5. Compute EOFs for January SST, T(2m), and SLP (Use `data(DNMI.xxx)`). Choose some locations from the NARP data, and use `DS()` to downscale the temperature from the three different predictors, respectively. Compare the results. Repeat for precipitation.

Chapter 8

REDUCING UNCERTAINTIES

It is crucial to know how well local climate statistics can be predicted before using it to make future climate scenarios. It is, therefore, necessary with an evaluation of both the global coupled general circulation model (GCM) results as well as the methodology that infers changes to local climates from the information in the GCM results.

Hence we will focus on the *optimal models* which are calibrated on a selection of empirical orthogonal functions (EOFs) that maximize the prediction skill. The purpose of developing these statistical models is to produce regional climate scenarios from given GCM results, and we will discuss the suitability of the ESD models for such applications.

8.1 Cascading Uncertainties

Climate change modeling for the future is associated with a number of different sources of uncertainty, starting with the fact that nobody knows what the external forcings will like in the future. The continuing emission of greenhouse gases may change as a result of changed human behavior, economics, politics, or access to resources. In addition, natural forcings such as volcanoes and solar activity may affect the future state of external forcing.

Then the GCMs themselves are not perfect, and unknown biases and errors constitute a significant source of uncertainty. Downscaling introduces further uncertainties because the ESD is non-perfect, GCMs are non-perfect, and the observed records may contain errors.

ESD may, on the other hand, also correct some systematic biases in the GCM, such as biases in the mean, but this depends on the ESD strategy.

Fig. 8.1. Living with uncertainties: although it is impossible to predict the exact shape and position of these condensation drops, it is nevertheless trivial to predict whether condensation drops form or not.

Since we are concerned with ESD here, we will look more closely into how we can reduce the uncertainties associated with downscaling.

We will also discuss how the calibration diagnostics can be used to provide a picture of how well the ESD models perform.

The error sources associated with conventional empirical downscaling based on EOF products can be expressed as the sum of GCM misrepresentation (systematic errors: e_{GCM}), errors due to mismatch between observed and simulated climatic patterns (observed EOFs not spanning the data space of the model results and sampling fluctuations in the observations: e_{pattern}), errors associated with the empirical model (linear approximation: e_{Ψ}), and natural variability (e_{noise}):

$$e_{\mathrm{tot}} = e_{\mathrm{GCM}} + e_{\mathrm{pattern}} + e_{\Psi} + e_{\mathrm{noise}}. \quad (8.1)$$

Although the second term, e_{pattern}, also relates to the systematic GCM errors, it has been separated from the former term and defined as the additional errors introduced by uncertainties related to the identification of observed spatial structures in the GCM model results.

In mathematical terms, EOFs of the observation may not necessarily span the data space of the model results. This may be true even if the model is perfect, as both observational record and model predictions are finite records, and the best EOF estimates of a finite time series may not be the same as the *true* EOFs because of sampling fluctuations (North *et al.*, 1982).

In the case of Benestad (1999c), e_{pattern} mainly comes from uncertainties associated with the regression of simulated climate patterns onto observed spatial climate structures, and the comparison between the common EOF and the EOF projection method suggests that the trend error is of the order 0.1°C/decade. The common EOF method eliminates the error term e_{pattern}.

Benestad (1999c) argued that by using best-fit trend estimates for long time intervals, the fluctuation errors associated with e_{noise} can be reduced to a minimum and systematic errors associated with the use of inappropriate control integrations can be eliminated.

The remaining errors can be mainly attributed to systematic GCM biases and shortcomings of the linear assumption that the temperatures were linearly related to the SLP fields, $\vec{y} = \Psi \vec{x}$.

8.2 De-Trending in the Calibration

The presence of trends in two different data trends will affect the fitting of the models. However, it is not guaranteed that the different trends are related to each other. De-trending refers to the procedure of removing these trends. The statistical models are then calibrated with the de-trended series, for which the degrees of freedom are high and the risk of coincidental false matching is much lower than between the two trends.

It is important to eliminate elements which can bias the results. Regression analysis aims to minimize the RMSE between two data series, and as a consequence, will always find a trend which gives the optimal RMSE fit between the two curves. A best-fit with non-zero trend may not necessarily be representative of a physical link if either quantity is a function of more than one factor.

The relationship between the two long-term trends may be due to coincidence, and can in worst case lead to invalid conclusions as shown in Benestad (2001a). Non-zero trends can also bias correlation analyses. Hence, best-fit analysis between two records that have non-zero trends may give a *biased* best-fit, and the time series should be de-trended prior to the analysis in order to obtain an *unbiased* best-fit.

If there is a real (and linear) relationship between the two quantities, then a regression model based on the de-trended series should also capture the relationship between their respective long-term trends. In simple mathematical terms, the time series can be expressed as the sum of a de-trended part and a linear trend: $x(t) = x_{\rm d}(t) + x_{\rm t}(t)$, where $x_{\rm t}(t)$ describes the linear trend in $x(t)$ and $x_{\rm d}(t)$ is the de-trended part. Hence the linear model $y(t) = ax(t)$ implies that $y_{\rm d}(t) + y_{\rm t}(t) = a(x_{\rm d}(t) + x_{\rm t}(t))$ and that the coefficient a is the same for the de-trended records and the linear trends.

It is important to only de-trend the calibration data, and not the data used to project into the future (no trend in the data will presumably give no trend in the ESD results).

Radan Huth argues that there is a paradox in the downscaling, as the models are calibrated with short-term variations and used to make predictions for long-term changes. He illustrates this by showing that two series may have a very high correlation, but different long-term behavior, such as trends (Fig. 8.2).

In such cases, statistical models may not capture the long-term variability, and one classical example of such shortcomings is where SLP

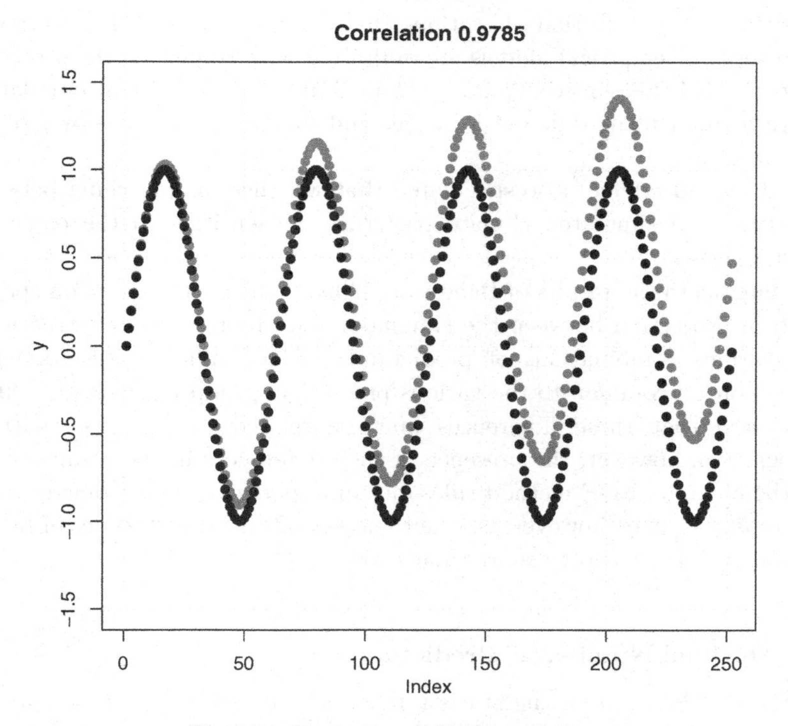

Fig. 8.2. Illustration of Huth's paradox.

is used to model the temperature (Hanssen-Bauer, 1999; Benestad, 2001a, 2002a).

By de-trending, the situation is not improved, and Huth's paradox may be used as an argument against de-trending. However, if we do not know *a priori* whether the trend is a part of the signal or caused by other irrelevant factors (a question of attribution), then the safest bet is still to focus on the short-term variations as they represent a higher degree of freedom and smaller chance of accidental good fit.

Furthermore, the issue of capturing the long-term behavior is addressed by the four assumptions discussed in Chapter 2, Sec. 2.2.1: in addition to a strong relationship, the predictors must be the representative of the local changes and must give a description of the change and the relationship must be stationary.

8.3 Using Different Predictands

Benestad (1999c) found that the downscaled future climate scenarios were sensitive to the selection of stations that was included in the predictand data set. The empirical models are optimized with respect to the predictor patterns and this sensitivity was explained in terms of different circulation patterns influencing different locations and that the model was over-fit for some stations.

Benestad (1999c) also speculated that whether the matching between observed and simulated climate patterns was sensitive to the choice of predictands.

Similar to the results of Benestad (1999c), which were based on spatial pattern projection between the simulated and observed model structures, the various combinations of predictands yielded different results for a given station common to the various predictand groups. Benestad (1999c) suggested that these differences may be related to imperfect pattern recognition, however, the presence of such differences in the results based on the common EOF method rules out an imperfect match. Some of these discrepancies may, however, be due to noise and the uncertainties of fitting a best-fit linear trend to short time series.

8.4 Optimal Number of Predictors

In the development of the statistical models, it is important to find the optimal number of predictands that yields the best prediction scores.

The number and type of predictors must be selected carefully in order to maximize the skill and avoid overfitting (Wilks, 1995, p. 185).

For instance, combinations of noise or signals unrelated to the predicted quantity may give a good fit to the data used in the training of the model, but will usually not produce good predictions.

One method to construct models with optimal skill and avoid overfit involves cross-validation and the use of a screening technique to estimate the optimal number of predictors.

8.5 Trends versus Time Slices

Natural decadal variations tend to be pronounced in some regions (e.g. northern Europe), and since these are regarded as chaotic, it is impossible to reproduce the exact time evolution. Figure 8.3 shows a spectrogram for the Oslo temperature, revealing pronounced variations on time scales of 10 years and longer. Thus, there is some uncertainty associated with the question whether a year is in warm or cold decade.

If multi-model ensembles are used, then this error can be reduced due to the fact that the time slices from each different model may be dominated by different state on decadal time scales, thus providing a larger statistical sample.

The term "linear trend" is in this context used to mean the long-term temporal evolution of a given quantity, and is estimated through the linear regression in time $(t) : \hat{y}(t) = c_0 + c_1 t$.

Machenhauer *et al.* (1998) argued that the *sampling errors* associated with selecting random 10-year to 30-year time slices for the MPI CTL were less than 2 K for temperatures and less than 40% for precipitation. However, as there were sampling errors in *both* CTL and the scenario integrations, it is necessary to combine the fluctuation errors of the individual data sets ($\Delta T \approx \sqrt{(2\,\mathrm{K})^2 + (2\,\mathrm{K})^2} \approx 3\,\mathrm{K}$).

Machenhauer *et al.* (1998) argued that the sampling fluctuations were smaller than the *systematic biases.*[1] Dynamical downscaling based on 10-year long time slices are prone to high sampling uncertainties, as the signal

---

[1]A systematic error was defined as a bias which is essentially independent of the sample (time slice) chosen. However, estimating *systematic errors* from 10- to 30-year long time slices may easily be contaminated by decadal and inter-decadal variability (Benestad *et al.*, 1999).

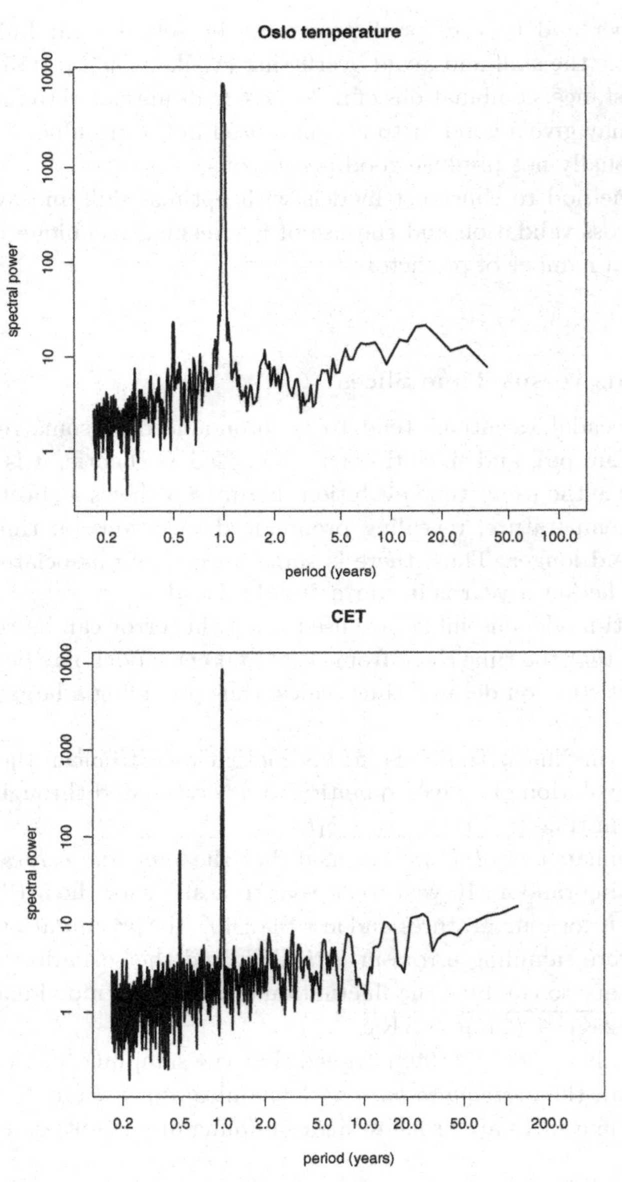

Fig. 8.3. Spectrograms for the monthly temperature in Oslo (top) and central England temperature (CET; bottom) show a strong annual cycle but also a degree of decadal variability.

magnitude is of the order of 1–3 K which is also similar to the sampling errors.

Benestad (2001b) suggested a more robust method for deducing climate change with empirical models by trend fitting (see e.g. Fig. 6.11). But trends are not always well-defined (Benestad *et al.*, 2007), and it only makes sense to talk about trends if they really provide a representative description of the behavior of the time series.

8.6 Domain Choices

Huth (2002) concluded that for the explained variance derived from a cross-validation, the size of the domain on which the predictors are defined plays a negligible role.

Benestad (2001b), on the other hand, demonstrated that the choice of domain may influence the trend estimates derived through ESD for temperature in Norway (Fig. 8.4). Moreover, when using a large domain which included a dipole pattern with both positive and negative corrrelations, Benestad (2002b) found that ESD could result in negative temperature trends despite a global warming.

The most recent versions of `clim.pact` include a function `objDS` that tries to minimize the errors associated with the domain choice, and automatically identifies a rectangular (in the longitude–latitude coordinate system) region on the basis of a correlation map between the predictor and predictand (Fig. 8.5). The algorithm then extracts the north–south and east–west profiles of the correlation structure, and the predictor region is then defined as where the correlation along these profiles is greater than zero (marked as vertical lines in the right panel in Fig. 8.5).

Smaller geographical predictor domains often yield the best cross-validation scores, which is according to the expectations. The highest scores, and hence the smaller predictor domains, were associated with the highest trend estimates.

Despite the higher scores, a small predictor domain size does not necessarily ensure a more realistic scenario, because small scale climatic features are expected to suffer more from GCM model errors than patterns with larger scales. Therefore, there is no way of telling whether the higher trend estimates are more realistic or due to model misrepresentation of the SLP over Scandinavia.

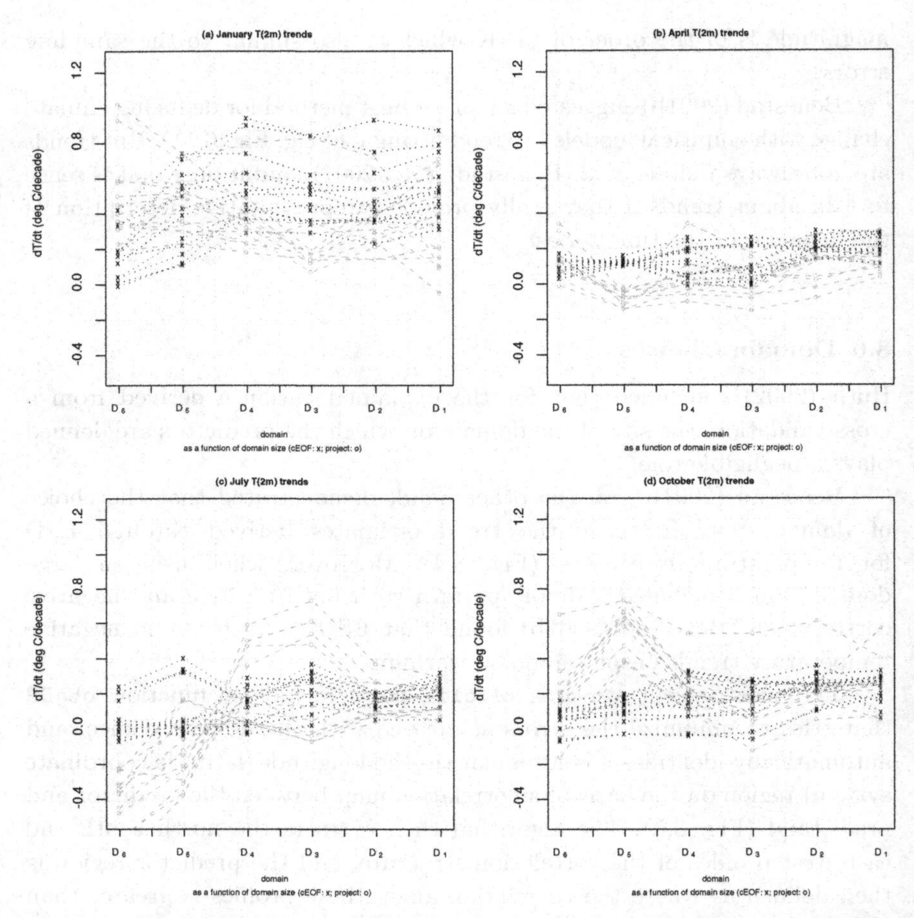

Fig. 8.4. Trend estimates as a function of the domain choice for January, April, July, and October. From Benestad (2001b). Gray represent "Perfect prog" and black common EOF-based approaches.

8.7 Spatial Coherence

After having compared ESD based on various regression- and CCA-based strategies, Huth (2002) concluded that the spatial correlations are reproduced most closely by CCA and the correspondence is best when more modes are included. He compared both regression with and without stepwise screening ("stepwise regression" and "full regression," respectively).

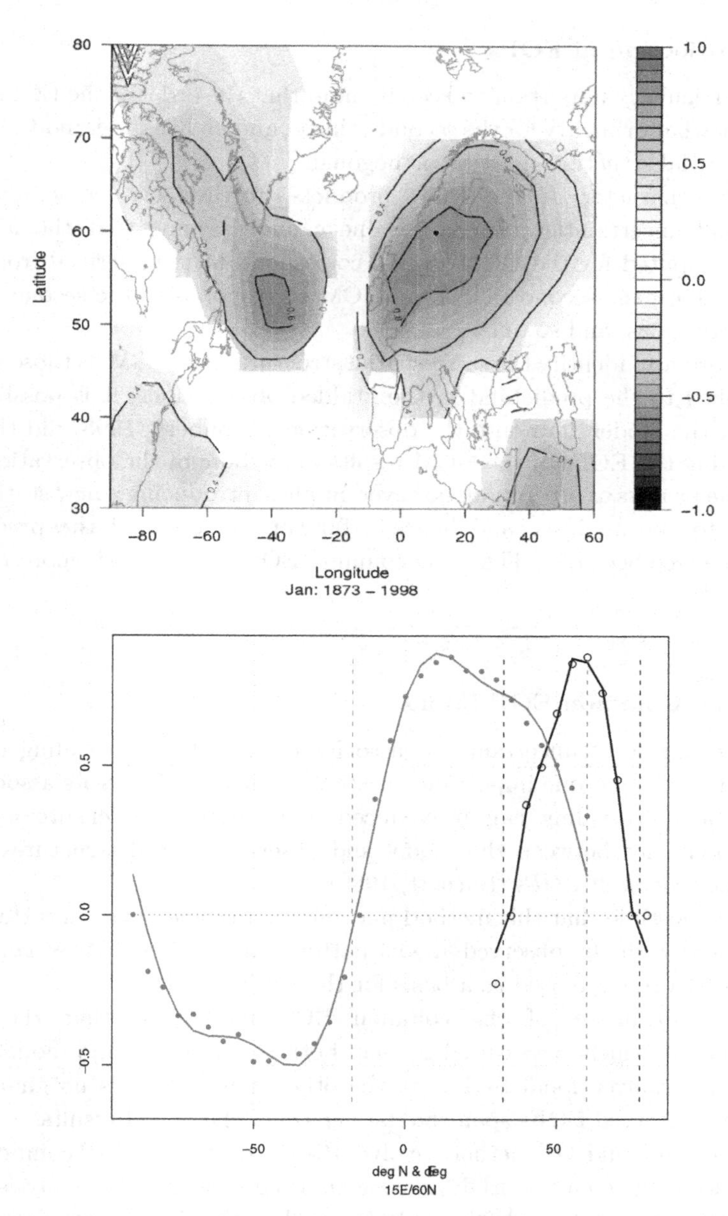

Fig. 8.5. Correlation map (top) and the north–south and east–west profiles (bottom) of the spatial correlation (along the faint yellow dotted lines in the bottom panel).

8.8 Projection of EOFs

It is particularly important to keep in mind that the order of the EOFs may be somewhat arbitrary for the second-, third-, and higher-order modes, since their spatial structure must be orthogonal.

The character of the EOF products, furthermore, is subject to sampling uncertainties. Therefore, one cannot assume that the leading EOF computed for the GCM results correspond to that derived from the observations, the second EOF from GCM corresponds to the second mode in observations, and so on.

In order to identify the same spatial structures in a GCM as those which are linked to the predictand in the gridded observations, it is possible to project the model data onto the observations. Schubert (1998) did this by projecting the EOFs of the GCM results onto those of the observations.

The process of projection, however, implies introducing a new statistical model to the analysis, and hence a further reduction of the predicted variance (see Sec. 7.1). Thus, the common EOF framework is more robust (Fig. 8.4).

8.9 The Common EOF Frame

A large degree of uncertainty is associated with the downscaling of the GCM results, and one important question is whether the errors associated with the downscaling can be reduced. One method to ensure a good correspondence between the model and observed spatial structures is to calculate *common EOFs* (Barnett, 1999).

It is possible that this method may eliminate some of the uncertainties associated with the observed–model pattern mismatches if these common EOF structures are used as a basis for the ESD models.

The advantage of the common EOF method is that they are eigenvectors which are assured to span both the observed and model data space. For conventional EOFs, on the other hand, there is no guarantee that the observed EOFs span the data space of the model results.

The common EOF method involves the use of the principal components (PCs) from the common EOF analysis in a regressional type analysis. The PCs from the common EOF analysis can be regarded as two parts: the observations and the model results.

As the PCs represent the temporal evolution of the same spatial patterns (modal structure) in both the observations and the model results,

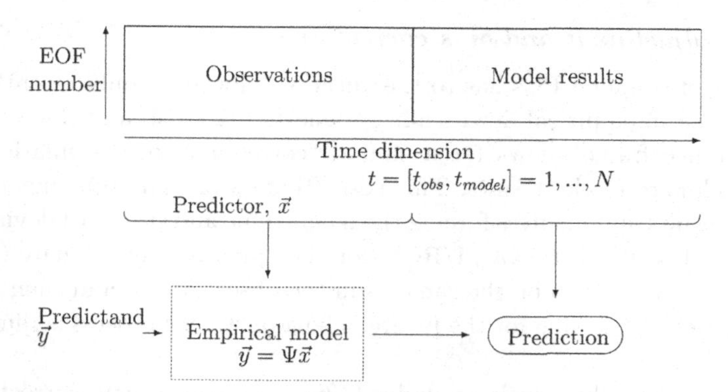

Fig. 8.6. A schematic illustration of the common EOF method. Here, y denotes the predictands (station observations) and x are the predictors taken from the common EOF PCs. First, the PCs corresponding to the observations and station data are used for model calibration, then the corresponding PCs from the GCM model results are used for predictions.

the empirical model obtained with the part of the PCs describing the observations can be applied directly to the model results (the remaining part of the PCs). Figure 8.6 illustrates the process of training the empirical model and then using the model in conjunction with the model PCs to make predictions.

There are various ways to apply common EOFs to downscaling, as there are with more conventional methods, and there is no clear *a priori* optimal method. The empirical models can be based on either a canonical correlation analysis (Benestad, 1998a, 1999c), singular value decomposition (SVD) (Benestad, 1998b) or multi-variate regression (MVR) (Benestad, 1999b). All these model types can be trained with the leading common EOFs through stepwise screening calibration (Kidson and Thompson, 1998; Wilks, 1995), in which the contribution of each PC is evaluated through a cross-validation analysis (Wilks, 1995).

Only those that contribute to the cross-validation skill are then included in the predictor data set. This approach reduces the risk of over-fitting the models and simultaneously extracts as much useful information as possible from the predictor data (optimizing e_Ψ).

Furthermore, the common EOF framework is also ideal for nonlinear downscaling, $\vec{y} = \Psi(\vec{x})$, such as neural networks and classification schemes. Together with estimating the long-term mean change by best-fit of linear trends for given long intervals, the common EOF method provides a sound strategy for future empirical downscaling.

8.9.1 *Adjustment and bias corrections*

The use of common PCs has to the authors' knowledge only recently been introduced in empirical downscaling (Benestad, 2001b), and this new type of reference frame allows for a simple "correction" of systematic biases in the climate model results. This correction entails an adjustment of the model results and involves forcing the mean value and standard deviation of the PCs describing the GCM/RCM for the "present-day" climate (control period, or "CTL") to be the same as in the observations, and then use the same offset and scaling for the future. The adjustment process is illustrated in Fig. 8.7.

For some fields, such as daily T(2m), an adjustment correcting for systematic model biases is required in order to obtain realistic distributions in the downscaled results. As described in the method section above, the adjustment may consist of setting the mean and standard deviation of the part of the PCs describing the CTL to the same values as those of the observations.

The adjustment forces the CTL to have similar features as those observed in terms of the location and spread of the PCs. The PCs describe the weighting of the EOFs (Lorenz, 1956; North *et al.*, 1982; Preisendorfer, 1988) for the combined data, and hence the common spatial climatic patterns for both observations and model. A large offset or a scaling factor substantially different from unity is an indication of substantial systematic bias in the GCM results. The offset and scaling factor can be used for comparing the model skill.

8.10 Further Reading

Benestad (2001b) found a greater sensitivity of the trend estimates on the choice of domain when the EOFs from the GCMs were projected onto those of the observations, than if a common EOF frame was used (Fig. 8.4).

Benestad (1999c) used global climate change simulations from coupled GCMs and empirical downscaling models in an attempt to predict changes to the local climate. Although the GCMs gave a good reproduction of the large scale geographical climate patterns, the matching of observed and simulated spatial patterns introduced additional uncertainties and the possibility of additional errors. Benestad (2001b) then evaluated a technique which aims to eliminate these additional error sources by using *common*

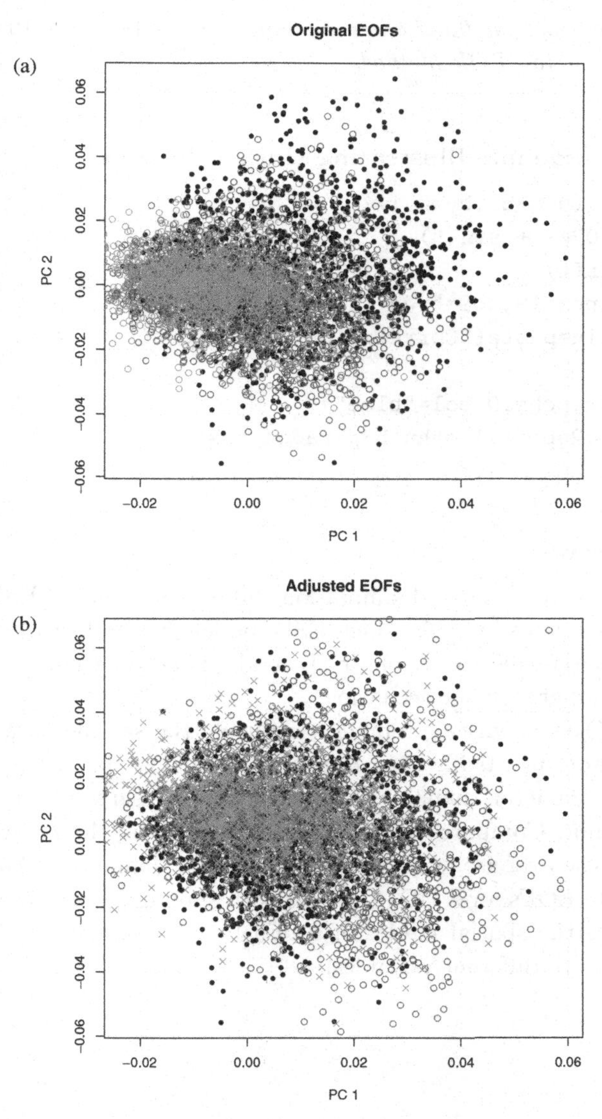

Fig. 8.7. An illustration of how the GCM results can be adjusted in order to ensure that the PCs describing the control-period (CTL; gray) have the same location and spread as the observations. The adjustment consisted of centering and scaling the CTL part of the PCs to match the observations (black), i.e. subtracting the mean value for the CTL part, multiplying with the fractional standard deviation ($s_{\text{obs}}/s_{\text{CTL}}$), and adding the mean for the part of the PCs representing observations. Panel (a) shows a scatter-plot between PCs 1 and 2 for the observations and the unadjusted results for the CTL, whereas (b) shows the adjusted PCs. The example shown here is taken from Imbert and Benestad (2005).

empirical orthogonal functions (*common EOFs*) (Barnett, 1999), referred to as the *common EOF method.*

8.11 The Example Illustrating Huth's Paradox

```
> t <- seq(0,8*pi,by=0.1)
> y <- 0.02*t + sin(t)
> x <- sin(t)
> plot(y,pch=19,col="red",ylim=c(-1.5,1.5),
       main=paste("Correlation",round(cor(x,y),4)))
> grid()
> points(x,pch=19,col="blue")
> dev.copy2eps(file="huth-paradox.eps")
```

8.12 Exercises

1. How can errors in GCM simulations influence results of ESD?
2. It is possible to identify which ESD models fit the historical data best. But is it possible to say one ESD model is better than another for the future climate change scenarios?
3. Use `DS()` to downscale temperature using the sample data provided in `clim.pact` and `DNMI.t2m`. Repeat the exercise by using successively smaller predictor domains. Compare with a regression against a grid point value. Compare the results (time series and the R^2-scores).
4. Repeat the exercise using different predictor domains and `DNMI.sst`.
5. Compute EOFs for a GCM and for corresponding gridded observations. Compare the spatial modes of EOFs 1–4. Comment on what you see. Repeat with different variables.

Chapter 9

DOWNSCALING EXTREMES AND PDFs

Climate and weather extremes have been the subject of study in the EU project STARDEX (http://www.cru.uea.ac.uk/cru/projects/stardex/) under the Fifth Framework Programme (2002–2005) as one of its overall objectives is to "provide scenarios of expected changes in the frequency and intensity of extreme events." The scientific objectives encompasses the identification of robust methods for inferring extreme events through downscaling and providing a standard set of indices describing extremes in Europe (Fig. 9.1).

Methodologies for statistical downscaling or modeling extremes are often invalid under changing conditions, but one approach involving computation of a set of indices or upper/lower percentiles may be one solution. Alternatively, the recurrence of records and trends in extreme-event counts or threshold analysis (Benestad and Chen, 2006) can provide useful diagnostics.

9.1 Outliers and Extremes

Real data often contain errors of various sorts. These may be due to instrumental failure, contamination, mis-typing, etc. Some data have real values which are significantly different to that of the bulk of the data. These extreme values are referred to as *outliers*, and may affect various statistics severely (i.e., the standard deviation).

Extreme climatic events may have severe impacts on the society, economy, and the ecosystems, and are often the focus of risk management. For instance, hydroelectric dams may sustain water up to a critical level: how often can one expect high rainfall to cause the water level to exceed this threshold?

Fig. 9.1. Waterfalls can be viewed as an extreme behavior in a river flow if one looks at the flow velocity, with completely different character than in the rest of the river. Yet the two aspects are related, as the through-flow of water mass in the waterfall must match that in any other part of the river, even the more quiescent parts.

Extremes often do not go unnoticed, as they may have a severe effect on the environment and the society. For instance, the nearby presence of a hurricane or a typhoon is hard to miss. Thus, extremes may often have independent verifications in terms of their implications, although it is harder to provide independent estimates of the exact magnitudes.

9.2 Robustness and Resistance

Sometimes "robustness" and "resistant" have different meanings (Wilks, 1995). "A resistant method is not unduly influenced by a small number of outliers, or wild data" (Wilks, 1995, p. 22). On the other hand, a robust result does not hinge on the exact method used to derive it, and a robust method performs reasonably well in most circumstances, but is often not optimal.

A *robust* method, according to Press *et al.* (1989), is insensitive to small departures from idealized assumptions for which an estimator is optimized. "small" can mean (i) fractionally small departures for all data points; (ii) fractionally large for a small number of data points (outliers). This is essentially the same meaning as Wilks (1995) "resistant method."

9.2.1 *Estimators*

Estimators are used to derive values for parameters used in statistical models or measures (e.g., models for the statistical distributions).

M-estimates: Follow from maximum-likelihood. For example, least-squares. Often assume that the data are normally distributed (often not a good assumption for daily rainfall).

L-estimates: Linear combinations of order statistics (i.e., median, quantiles).

R-estimates: Based on rank *tests*, i.e. the inequality of two distributions can be estimated by the Wilcoxon test.

9.3 Probability Density Functions

Before proceeding we should stop and think about what probability density functions (PDFs) represent. To illustrate the use of PDFs, let us pretend to have a record of some climate variable and want to know: How often do we get months which are hotter than 30°C? Or: How often do droughts recur? In order to answer this, one may look at the distribution of the temperatures or rainfall and count how many times the events have happened before.

9.3.1 *What is probability distribution and why use it?*

If we know the likelihood (probability, chance, risk, p) that an event, E_1, will take place, then we can make some predictions for the future. For instance, if the probability of getting a warm spring in western Norway is $p = 0.64$ after having observed warm sea surface temperatures in the North Sea during the preceding winter, one may use this knowledge to make seasonal predictions. In the long run, such predictions should give more (correct) hits than a pure guess ($p = 50\%$).

Probability distributions and density functions describe the observed or expected frequency that a value has taken/will occur.

Conditional probability: $\Pr(E_2|E_1) = \Pr(E_1 \text{ and } E_2)/\Pr(E_1)$, the probability that E_2 will happen *given that* E_1 has occurred (here "Pr" represents the probability function and "p" a given probability). This is equivalent to Bayes' theorem (Leroy, 1998), more commonly expressed as $\Pr(X|Y) = \frac{\Pr(Y|X)\Pr(X)}{\Pr(Y)}$. Whereas time series of a variable shows the chronological evolution, a distribution plot does not provide any information about the chronology but shows how often a particular value has been observed and is a kind of *histogram*. A normalized theoretical probability distribution is also known as a *probability density function* (PDF). A distribution curve is useful for estimating the probabilities associated with certain events. For instance, to estimate the probability that a temperature is lower than $T_{\rm low}$, then $\Pr(T < T_{\rm low})$ = area of the part of the curve that corresponds to temperatures lower than $T_{\rm low}$ divided by the total area of the curve. To estimate the probability of seeing warmer months than $T_{\rm hot}$: $\Pr(T > T_{\rm hot}) = 1 - \Pr(T \leq T_{\rm hot})$.

There are two types of distributions: (i) continuous distribution such as Gaussian, exponential, Gamma, Weibull, etc.; and (b) discrete distributions (Poisson, Binomial). The latter can only describe integer values.

9.3.2 *Normal/Gaussian distribution*

$$f(x) = \frac{1}{\sigma\sqrt{2\pi}} \exp\left[-\frac{(x-\mu)^2}{2\sigma^2}\right]. \tag{9.1}$$

There are a number of commonly used theoretical distribution functions, which have been derived for ideal conditions. One such case is where the process $(y_i, i = [1 \cdots N])$ is random (stochastic), and whose distribution follows a Gaussian shape described by $f(x)$ in Eq. (9.1).

This distribution function is widely used in statistical sciences, where σ in this case is estimated by taking the standard deviation: $\sigma = \mathrm{std}(\vec{x})$, and μ is taken as the mean value of $\vec{x}$.

Figure 9.2 shows a typical example of a Gaussian distribution. The values of σ and μ have been taken from the Bergen September 2-m temperature 1861–1997 record, and the empirical histogram for the temperature record is also shown as black dots.

The Gaussian distribution function in Fig. 9.2 gives a concise and approximate description of the Bergen September temperature range and likelihood of occurrence. The mean and standard deviation, the two parameters used for fitting the Gaussian function to the observations, give a good description of the Bergen temperature statistics.

Gaussian distribution is also commonly referred to as "normal distribution."

One important property of the Gaussian distribution is the fact that the *central limit theorem* applies: as the sample size of a set of independent observations becomes large, the sum will have a Gaussian distribution.

It is possible to use ESD to predict both σ and μ independently over predefined time intervals, and then use these to reconstruct variations in the PDF.

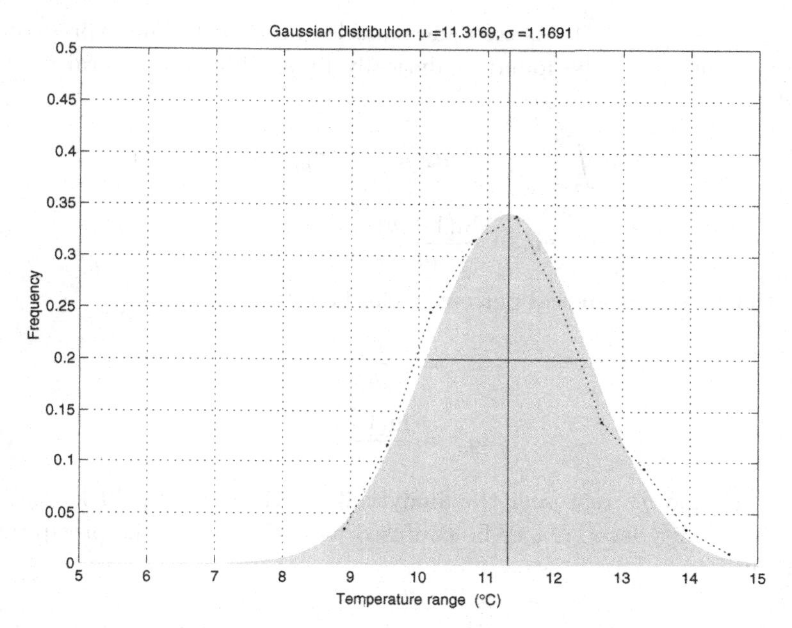

Fig. 9.2. An example of a Gaussian distribution curve. The vertical line marks the mean value and the horizontal line shows the $\pm\sigma$ range. The empirical probability distribution for the Bergen September temperature is also shown as black dots.

Sometimes, the Gaussian distribution curve is not a good description of the data because these are not symmetrically distributed with respect to their values. In such cases, non-symmetric distribution functions may be used to describe the data, such as the exponential or gamma distributions.

9.3.3 *The exponential distribution*

The PDF for the exponential distribution can be written as $f(P_{\rm w}) = -me^{mP_{\rm w}}$ if $m < 0$ because the area under the PDF curve must equal unity. It is also possible to derive an analytical solution for the mean of the PDF (often the "rainfall intensity — the mean of the rainy days"):

$$\begin{aligned}
\bar{P}_{\rm w} &= \int_{x=0}^{\infty} -mx\mathrm{e}^{mx}dx \\
&= -m\left(\left[\frac{x}{m}\mathrm{e}^{mx}\right]_0^{\infty} - \int_{x=0}^{\infty}\frac{1}{m}\mathrm{e}^{mx}dx\right) \\
&= m\left[\frac{\mathrm{e}^{mx}}{m^2}\right]_0^{\infty} \qquad \therefore \bar{P}_{\rm w} = -\frac{1}{m} \quad m < 0. \qquad (9.2)
\end{aligned}$$

The term $[\frac{x}{m}e^{mx}]_0^\infty$ is zero and cancels and $e^0 = 1$. The expression for the percentiles can be found analytically by solving the integral over the PDF:

$$p = \int_{x=0}^{q_p} -me^{mx}dx = [-e^{mx}]_0^{q_p} = -e^{mq_p} + 1$$

$$m < 0 \rightarrow q_p = \frac{\ln(1-p)}{m}. \tag{9.3}$$

Hence, the mean and percentile can be estimated by

$$\mu^+ = -\frac{1}{m},$$

$$q_p^+ = \frac{\ln(1-p)}{m}. \tag{9.4}$$

Here, μ^+ and q_p^+ represent the analytical solutions for Eq. (9.4), and p is the probability level, not to be confused with P which is the precipitation amount.

9.3.4 *Gamma distribution*

$$f(x) = \frac{(x/\beta)^{\alpha-1}\exp[-x/\beta]}{\beta\Gamma(\alpha)}, \quad x, \alpha, \beta > 0. \tag{9.5}$$

There are different ways of estimating the two parameters, of which the moment estimator (Wilks, 1995, p. 89) is the simplest one:

$$\hat{\alpha} = \frac{(\bar{x}_\mathrm{R})^2}{s^2}, \tag{9.6}$$

$$\hat{\beta} = \frac{s^2}{\bar{x}_\mathrm{R}}, \tag{9.7}$$

where $\bar{x}_\mathrm{R}$ is the mean value for the rainy days (here, day with rainfall greater than 1 mm) only and s corresponding standard deviation. These should not be confused with the more traditional "seasonal means" ($\bar{x}$) which are estimated for the entire season (both dry and rainy days).

From the estimated "shape parameter" $\hat{\alpha} = \frac{(\bar{x}_\mathrm{R})^2}{s^2}$ (Note: $\alpha \neq \frac{\overline{(x_\mathrm{R}^2)}}{s^2}$; (Wilks, 1995, p. 85) may give the impression that the estimator uses the mean of the square) and the "scale parameter" $\hat{\beta} = \frac{s^2}{\bar{x}_\mathrm{R}}$, we can get $\hat{\alpha}\hat{\beta} = \bar{x}_\mathrm{R}$. Another method to estimate the gamma-parameters is the *maximum-likelihood* fitting described by Wilks (1995) on p. 89, but this method does not allow negative and zero values (can be avoided by using only non-zero values).

$\Gamma(\alpha)$ denotes the *gamma*-function defined as

$$\Gamma(\alpha) = \int_0^\infty t^{\alpha-1} e^{-t} dt.$$

The gamma function has a useful property, which is

$$\Gamma(\alpha + 1) = \alpha\Gamma(\alpha). \quad (9.8)$$

If $\Gamma(\alpha)$ is known for any value $\alpha < 1$, then it is easy to calculate the corresponding value for and the number with similar decimal points.

Figure 9.3 shows the distribution function for daily precipitation in Oslo between 1883 and 1964.

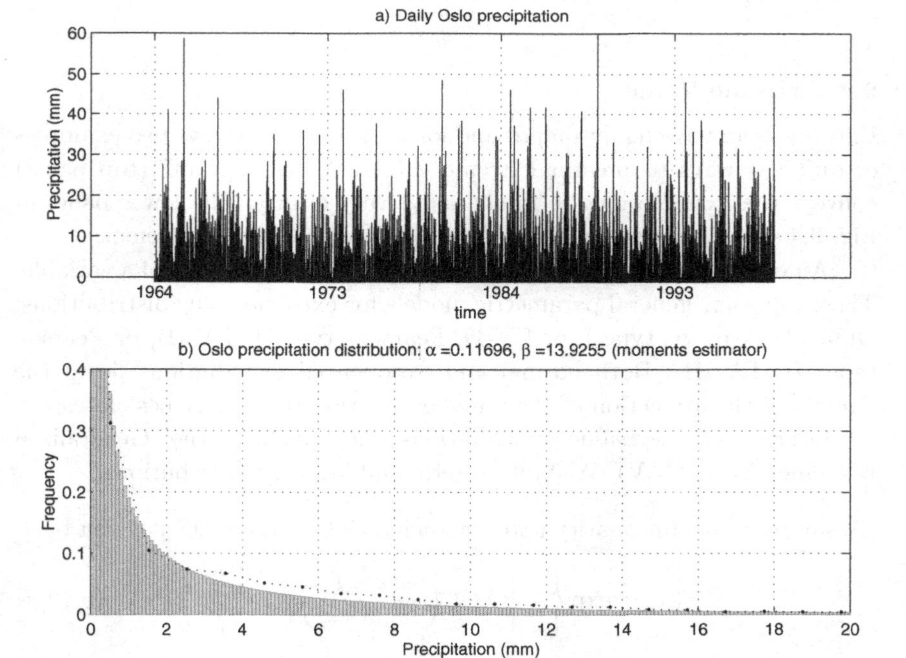

Fig. 9.3. An example of a non-symmetric distribution and the best-fit gamma distribution. The plot is based on the daily precipitation measurements made in Oslo (St. Hanshaugen), 01-Jan-1883 to 31-Jul-1964, after which the measurements were made at Blindern.

Wilks (1995) states that the moment estimators are "inefficient" and lead to erratic estimates, and recommends using the so-called maximum likelihood estimators:

$$\hat{\alpha} = \frac{1 + \sqrt{1 + 4D/3}}{4D}, \tag{9.9}$$

$$\hat{\beta} = \frac{\bar{x}_{\mathrm{R}}}{\hat{\alpha}}, \tag{9.10}$$

$$D = \ln(\bar{x}_{\mathrm{R}}) - \frac{1}{n}\sum_{i=1}^{n} \ln(x_i). \tag{9.11}$$

The central question here is whether the gamma parameters for a given location are systematically influenced by either the large-scale conditions or the local geography in such a way that they can be predicted given this information. Such predictions are in essence of the same as "downscaling."

An alternative to the gamma distribution is the Weibull function.

9.4 Extreme Values

Extremes can have many shapes and forms. Figure 9.4 shows two examples of different ways to present extreme values. A "cumugram" (top panel) conveys the extreme long-term character very clearly. The lower panel in Fig. 9.4 shows a more conventional way of presenting the extremes.

An extreme value is the largest or smallest observed value of a variable. There are some general parametric models for extreme value distributions: Gumbel (Pearsons type I, or EV-1), Pearsons type II (EV-II), or Pearson type III (EV-III). Both normal and exponential distributions lie in the domain of the attraction of the Gumbel function (the latter lies closest).

Other extreme value distributions may include the Generalized Extreme Value (GEV), Weibull, Pareto, and Wakeby distributions.

Assumption of stationarity and ergodicity.[a] The GEV CDF is given by

$$F(x) = \exp\left\{ -\left[1 + \xi\left(\frac{-(x-\xi)}{\beta}\right)\right]^{-1/\xi} \right\}, \tag{9.12}$$

[a]Every trajectory will eventually visit all parts of phase space and that sampling in time is equivalent to sampling different paths through phase space (von Storch and Zwiers, 1999, p. 29).

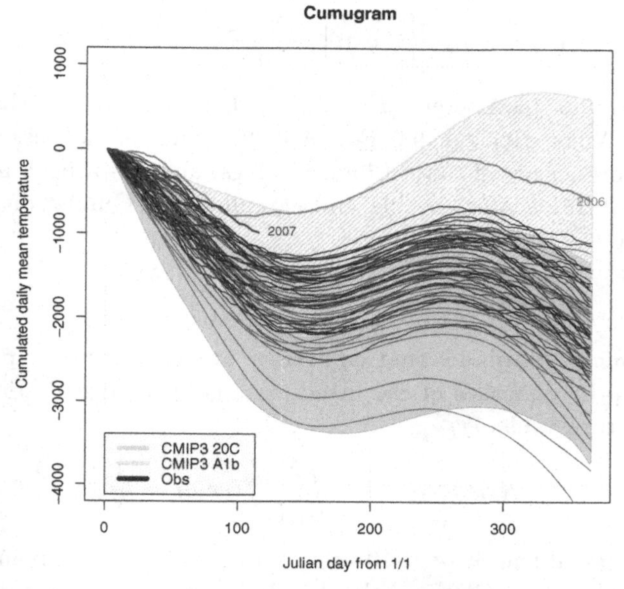

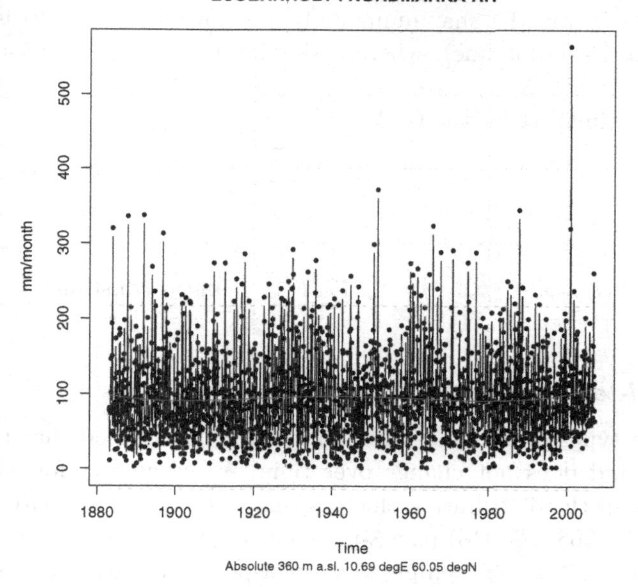

Fig. 9.4. Upper: A "cumugram" showing the cumulative temperature for Svalbard over the course of the seasons. The lines show actual measurements, whereas the gray and pink hatched areas show ESD results based on Benestad (2005). Year 2006 is marked by the red curve, and was an extreme year. Lower: Time series plot of the monthly precipitation measured at Bjørnholt, near Oslo, Norway. The rainfall in November 2000 was extreme.

$$f(x) = \frac{1}{\beta} \exp\left\{ - \exp\left[\frac{-(x-\xi)}{\beta} \right] - \frac{(x-\xi)}{\beta} \right\}, \tag{9.13}$$

ξ = location parameter, β = scale parameter (NB, there is a typo in (Wilks, 1995, p. 98, Eq. (4.42)): a "minus" is missing). The distribution is skewed toward higher values, and has a peak at $x = \xi$. Equation (9.13) is integrateable, and the cumulative Gumbel distribution function is

$$F(x) = \exp\left\{ - \exp\left[- \frac{-(x-\xi)}{\beta} \right] \right\}. \tag{9.14}$$

Return values: thresholds that on average are exceeded once per *return period*. Upper quantiles of the fitted extreme value distribution. For a X-year return value, $rr_{(X)}$,

$$P(rr > rr_{(X)}) = \int_{rr_{(X)}}^{\infty} f(rr) dr = \frac{1}{X}. \tag{9.15}$$

The X-interval-length (e.g. 10-year) return value is the point on the abscissa where the CDF equals $(1 - 1/X)$. For example, the 10-year return value for the maximum daily precipitation in Oslo is 46 mm (shown as dash-dot line), whereas the 100-year return value is 63 mm (dashed).

The estimators for the GEVs are

$$\hat{\beta} = \frac{s\sqrt{6}}{\pi}.$$

and

$$\hat{\xi} = \bar{x} - \gamma\hat{\beta}, \quad \gamma = 0.57721\ldots \text{ (Euler's constant)}.$$

9.4.1 *iid-test*

For many types of extreme value analyses, it is assumed that the PDF is constant and does not change over time. A simple test for whether the upper tail of the distribution changes can be implemented with an iid-test (Benestad, 2003d, 2004d) (the R-package `idd.test`).

The iid-test is very simple and yet a powerful means to test whether the upper tail of the PDF if being shifted in time. Too many or too few record-events are indicators of a non-stationary PDF, as a H_0 corresponding to a constant distribution (i.e., the data being "identically distributed") implies a simple rule for how often we can expect to see new records due to sampling fluctuations.

The iid-test is based on the assumption that the probability that the nth element contains the greatest value is $p = 1/n$, and can indicate whether the series fails to either contain independent data or that they are not identically distributed. The test also fails if there are ties (Benestad, 2004d).

An iid-test will indicate whether an extreme value analysis, for instance based on GEV modeling, is appropriate. The test was originally designed to test the assumptions for using analog models, and it useful for judging the likelihood that the upper tail derived through analog models will be convoluted. Thus, the null-hypothesis of the iid-test is the basis for the statement that the analog models are flawed when it comes to modeling the upper tails during an ongoing climate change.

If the number of record events n_r is low for both "forward" and "backward" tests, then this may be a sign of ties or that the instrument "clips" the data (there is an artificial upper limit to the readings imposed by the instrument) (Fig. 9.5).

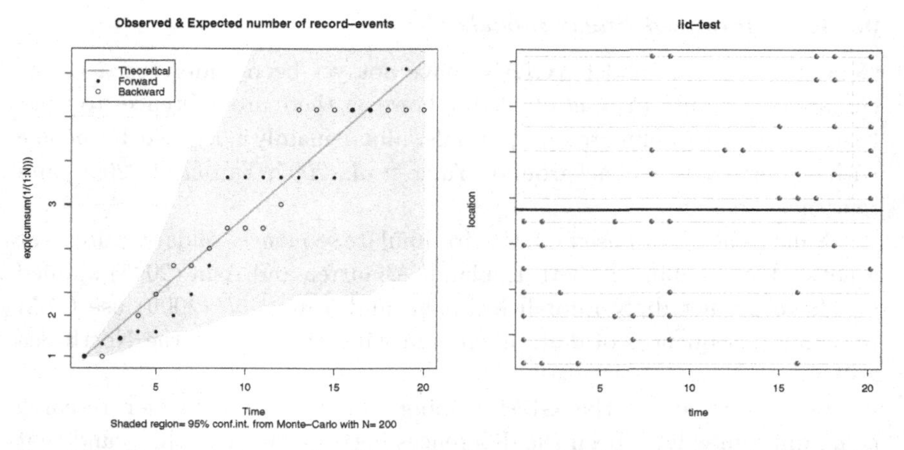

Fig. 9.5. Demonstration of the iid-test. Left panel shows how many record-breaking events n_r have been recorded from time zero as more and more observations are collected. Linear line is the expected number of n_r and the gray envelope marks the CI. Filled symbols show n_r for the chronological ordered data "forward test" whereas open symbols represent the analysis applied to a reversed time series "backward test". Right panel shows the timing of new record breaking events for parallel measurements made at a number of locations, both for the chronological ordered data and the reverse chronological order.

9.5 Downscaling Extreme Indices

It is possible to define a number of climate extreme indices (Alexander *et al.*, in press; Frich *et al.*, 2002), and then apply empirical–statistical downscaling (ESD) directly to those. Such work has been carried out within the European Union STARDEX-project.[1]

Frich *et al.* (2002) used an index (R95T) that described the fraction of the total precipitation associated with events exceeding the 95%-percentile. On a similar vein, Ferro *et al.* (2005) presented simple non-parametric methods for exploring and comparing differences in pairs of PDFs and histograms.

Extreme heat waves can be linked with the atmospheric circulation regime (Kyselý, 2002a), and a catalogue of weather type or some classification may be used to study their occurrence.

Often extreme indices may not follow a Gaussian distribution, and therefore an ordinary linear model, such as ordinary least squares, may not be the most appropriate means for downscaling. Methods such as CCA also assume a normal distribution. However, generalized linear models (GLM) can be employed for data with PDFs other than Gaussian.

9.5.1 *Generalized linear models*

Generalized linear models (GLMs) have not yet been widely used in the climate community (Yan *et al.*, 2006), however there are references to work based on GLM in climatology journals, albeit mainly authored by people with strong statistics background (Yang *et al.*, 2005; Yan *et al.*, 2006, and references therein).

Yang *et al.* (2005) used GLMs[2] to simulate sequences of daily rainfall at a network of sites in southern England. Abaurrea and Asín (2005) applied GLMs to predict daily rainfall in Spain, and Yan *et al.* (2006) use GLM to simulate sequences of daily maximum wind speeds over the North Sea region.

One obstacle for the GLM gaining impasse to the wider research community may have been the differences between the disciplines and that the papers on GLM have been written by statisticians with a rigorous formal statistical treatment that is often difficult to digest for climatologists with a physics, meteorology, or geography background.

---
[1] http://www.cru.uea.ac.uk/projects/stardex/
[2] http://www.statsoft.com/textbook/stglm.html

For GLM-based analysis, a common view is that each data value is one realization of a stochastic process, in contrast to a classical Newtonian physics view where the variable is regarded more as a deterministic response to a set of forcing conditions. We will refer to the former mind set as "Bayesian frame." Moreover, a GLM approach often assumes that the PDF is varying from time to time, as opposed to the PDF being constant over time but where each realization are predicted deterministically, given a set of predictors.

Although the differences between the statistician's frame of mind and that of a "typical physicist" may seem subtle, the implications are more profound since the PDF changes from one observation to the next in the former, and it is the PDF that is systematically affected by the forcing (Yan *et al.*, 2006). From a Newtonian view point, on the other hand, systems are often regarded as well-defined for which energy, momentum, and mass are conserved quantities. Ideally, any state can be determined accurately in this frame work if all the forcings are known, given these constraints.

Unknown factors produce behavior that is unaccounted for, commonly referred to as "noise." If the noise is weak (high signal-to-noise ratio), then the system is practically deterministic, but if the noise if overwhelming, then the system behaves in a stochastic manner.

Thus, from a Newtonian view point, the noise is not affected by the known forcing conditions and the PDF for the response is taken to be independent of that of the noise, whereas in the Bayesian frame, the PDF describing some stochastic behavior is assumed to be systematically influenced by the external conditions.

The implications of stochastic systems is that the principal conditions of causation are not accounted for because different realizations can be drawn from the same PDF, but there is no way to predict the exact value even if the PDF is well described.

In practical terms, however, the two approaches are two sides of the same problem, but involve completely different interpretations.

9.5.1.1 *Maximum likelihood estimation*

A maximum likelihood estimation (MLE) is used to derive a model for the expectation value $E(\cdot)$ for a variable Y_t at any given time t: $E(Y_t) = \mu_t = g^{-1}(X\beta)$. Here μ_t is the mean for the PDF ($\mu_t = \int_{-\infty}^{\infty} x f_t(x)dx$) at time t, rather than the empirical average for the entire sequence ($\bar{Y}$), and $g(\cdot)$ is a link function that determines the scale on which the predictors are additive.

Note that in a Bayesian framework, the GLM predicts the mean μ of the distribution rather than the realization itself. One example, where the mean daily precipitation is modeled using a GLM taking $\vec{x}_1$ and $\vec{x}_2$ as predictors, looks like

$$\ln(\vec{\mu}) = \beta_0 + \beta_1\vec{x}_1 + \beta_2\vec{x}_2. \tag{9.16}$$

This exercise can also be repeated with μ replaced by q_p for the prediction of percentiles ($\hat{q}_p$). The regression analysis just by itself, however, does not provide a description of the PDF, but if the estimates of μ are used in connection with equation describing the relationship between the parameter and μ (such as for an exponential distribution) then it is possible to use GLM to estimate PDFs too.

This approach is similar to that of Yan *et al.* (2006), who assumed either gamma or Weibull distribution and constant shape parameters, and then used μ to infer the PDF for a given time.

It is possible to combine direct methods (e.g., linear or analog) with the downscaling of PDFs to get time series with more realistic characteristics (variance and time structure). The downscaled PDFs can then be used in conjunction with local quantile transformation to ensure that the projected results follow a prescribed distribution (see example below).

9.6 Downscaling PDFs

When downscaling the PDF for a variable X, given external conditions Y, one implicitly employs Bayes' theorem (Leroy, 1998):

$$\Pr(X|Y) = \frac{\Pr(Y|X)\,\Pr(X)}{\Pr(Y)}. \tag{9.17}$$

Downscaling of PDFs is thought to provide a more realistic representation of the upper tails of the distribution than a direct downscaling of the daily values from the climate models (Hayhoe *et al.*, 2004; Pryor *et al.*, 2005, 2006).

Yan *et al.* (2006) assumed either gamma or Weibull distribution and constant shape parameters, and then used μ to infer the PDF for a given time. However, their work did not really concern ESD as such.

There are two ways to downscale PDFs, either to model the PDF parameters directly from the large-scale predictors, or to relate the PDF parameters to the local climatic conditions, and then downscale the latter before employing a statistical model to derive the parameters from the local climate information.

9.6.1 *Downscaling PDFs for temperature*

The simplest way of downscaling PDFs can be demonstrated for the temperature which is close to Gaussian. Figure 9.6 shows the results where the mean (estimator for μ) and standard deviation (estimator for σ) for each January months have been used as predictand in a two-tiered approach, where monthly mean temperature has been used as predictor for each.

Knowing the relationship between μ and σ on the one hand and the trends described by the large-scale predictor on the other hand, enables us to infer projections for the two PDF parameters.

The example given in Fig. 9.6 suggests that there is no trend in σ, but a substantial systematic increase in μ.

9.6.2 *Downscaling PDFs for daily precipitation*

The gamma parameters shape and scale also show a relationship with local climatic conditions (Benestad *et al.*, 2005), even though the relationship may be weak.

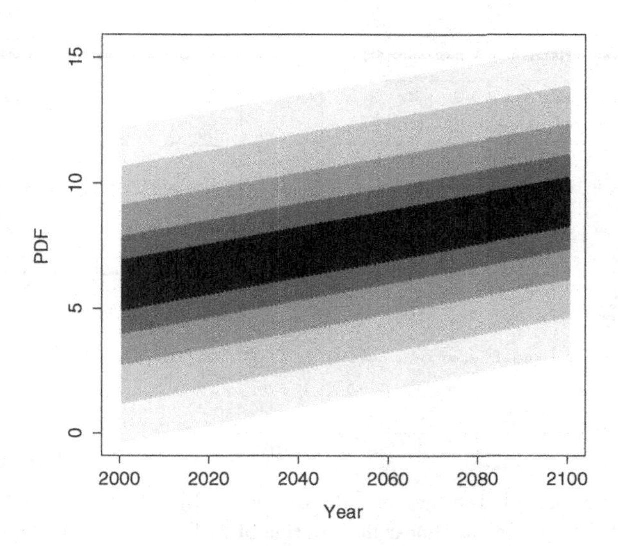

Fig. 9.6. Downscaled distribution for daily January temperature in Oslo.

Exponential distribution furthermore provides simple analytical solutions, and reduce the number of unknown parameters to fit. Furthermore, the upper tails involve a smaller number of events and are thus affected by uncertainty associated with statistical fluctuations to a higher degree than less extreme values.

Figure 9.7 is taken from Benestad *et al.* (2005) and shows a number of linear-log plots for the distribution of 24-h precipitation amounts for 49 different European locations. The gray dots represent the empirical results

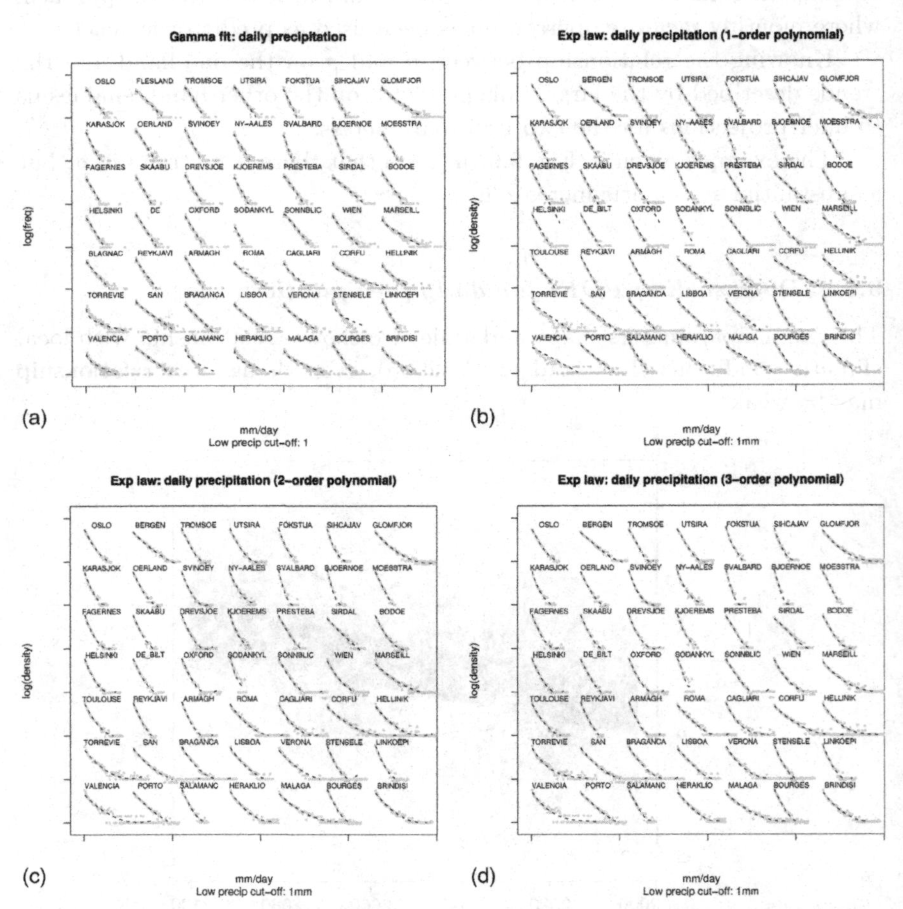

Fig. 9.7. Fits of (a) the gamma distribution, (b) e^{a+bx}, (c) $e^{a+bx+cx^2}$, and (d) $e^{a+bx+cx^2+dx^3}$ to the log-linear distribution of 24-h precipitation. The y-axis shows the log of the frequency and the x-axis shows the linear scale of precipitation amounts.

(histogram with log y-axis) and the dashed black lines show the distribution model fit. The different distribution models include three exponential models as well as the gamma model (using moments estimators).

When downscaling the precipitation, it is common to split the data between rainy days and dry days. Here, P will denote the precipitation in general terms, $P_{\rm w}$ will denote the precipitation for wet days, and $P_{\rm a}$ represents precipitation for all days, and the mean value of P is written as $\bar{P}$. Thus, $\bar{P}_{\rm w} > \bar{P}_{\rm a}$ since $P_{\rm a}$ includes a number of data points with zero value and $P \geq 0\,\text{mm/day}$.

All seasons may be included in the estimations of the PDFs and the mean climatic conditions ($\bar{T}$ and $\bar{P}_{\rm a}$), and the PDFs represent the local (single station) precipitation rather than areal means, and all seasons rather than a particular time of year.

When histograms for daily precipitation are plotted with a log-scale along the vertical axis, a close to linear dependency to magnitude is evident (Figs. 9.7 and 9.8).

If we let $y(P_{\rm w}) = \ln|\hat{n}(P_{\rm w})|$, and $\hat{n}(P_{\rm w})$ be the number of $P_{\rm w}$ events with values falling within the interval $P_{\rm w}$ and $P_{\rm w} + \delta P_{\rm w}$ ($\delta P_{\rm w}$ was taken to be 2 mm), then a linear model $\hat{y} = m' P_{\rm w} + c'$ can be used to represent the linear dependency between counts and magnitude.

A (weighted) least-squares regression can be used to solve for m' and c', taking $P_{\rm w}$ as the predictor and the log of the counts as the predictand (the weights can be taken as $\sqrt{n}$ in order to emphasis the cases with a greater statistical sample).

A linear dependency $\ln|n_P| \propto P_{\rm w}$ implies a simple exponential distribution, $n_P \propto e^{mP_{\rm w}}$ with $m < 0$, and has one advantage that the PDF can be written as $f(P_{\rm w}) = -me^{mP_{\rm w}}$ because the area under the PDF curve must equal unity. Another useful property is that the mean μ and any percentile q_p for the wet-day distribution can easily be derived analytically.

It is evident from Fig. 9.7 that the character of the distribution (e.g., the slope m) varies from place to place. The interesting question here is whether there is a systematic dependency between the slope m and the large-scale flow regime or dominant characteristics of the local climate.

Similar near-linear behavior can be seen in similar log-linear distribution of tornadoes of category F1 and greater (Feuerstein *et al.*, 2005, Fig. 2a). Feddersen and Andersen (2005) have used exponential distribution to approximate the PDF for 24-h precipitation in Denmark.

It is also possible to apply the regression results for the prediction of temporal changes. Figure 9.9 shows the prediction (extrapolation) of

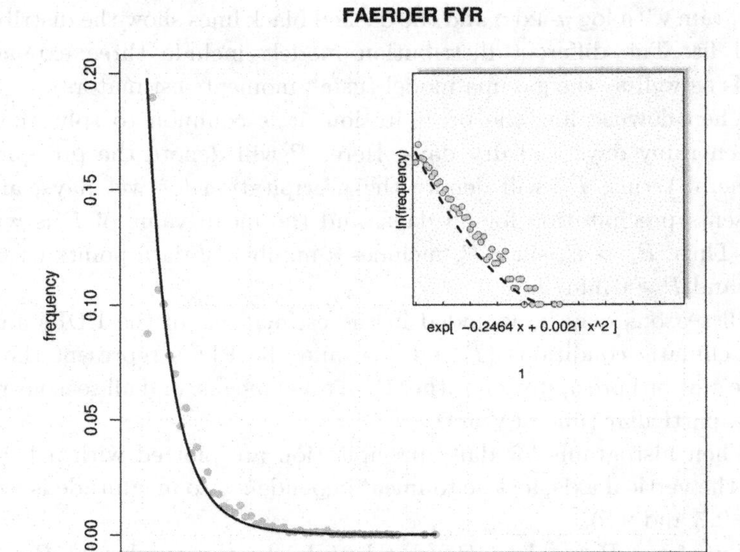

Fig. 9.8. The daily rainfall amount at Ferder and fitted exponential distributions.

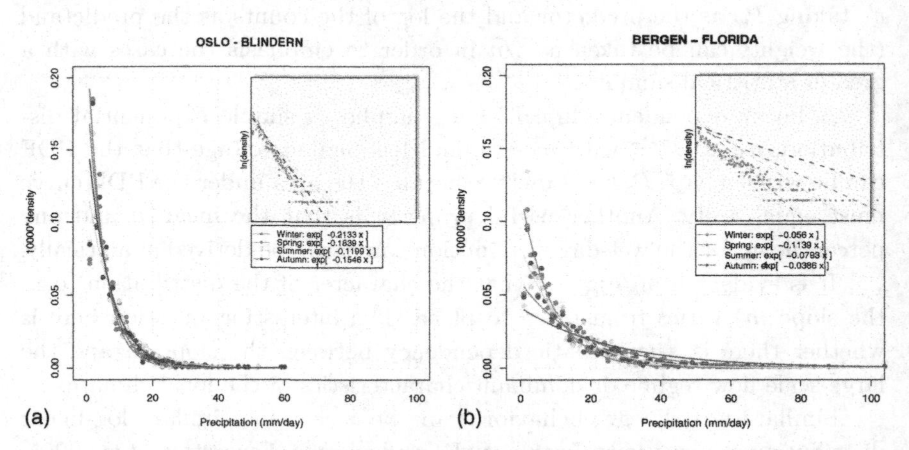

Fig. 9.9. Seasonal dependency: lines represent estimates for $f(x)$ and symbols represent the empirical values. The different seasons are shown in different colors.

variation of the seasonal 24-h precipitation distribution in Oslo and Bergen with a second-order polynomial and linear exponential models, respectively.

The model underestimates the frequency of days with low precipitation in Bergen during winter and autumn in this example, but yields an approximate representation of the distribution functions for Oslo.

Part of the discrepancy between the empirical and extrapolated representation is associated with the constant value not used in the estimation of the PDF. Here, the probability densities are used along the y-axis, and since these are significantly less than unity small errors tend to appear more serious than a linear plot would indicate. The extrapolated PDFs in the main frame show a better correspondence with the empirical data. Especially the curve for the autumn in Bergen (gray) indicates too low occurrence of drizzle and too many cases with heavy precipitation.

An extrapolation of the exponential law was used by Benestad *et al.* (2005) and Benestad (2007) to make projections for the future. The linear rate of change (°C per decade and mm/month per decade) in the local mean temperature and precipitation was taken from the downscaling analysis of Benestad (2005).

It is possible to extend the model by using a polynomial rather than a linear fit to describe the data points in the log-histogram. In some locations such as Oslo and Tromsø the results indicate little changes, whereas in places like Bergen and Glomfjord, the extrapolation indicates a substantial increase the frequency of heavy 24-h precipitation.

However, it is important to keep in mind the fact that these results may be subject to biases and that the validations using independent data suggest that these extrapolations are not always accurate for all locations.

The projections for Bergen and Glomfjord with a decrease in the days with drizzle and more dry days are considered not reliable since the physical situation for Bergen and Glomfjord with a dominance of orographically forced rainfall differs much from most other places, and hence an extrapolation based on other types of conditions (with a warmer climate but where the orographic effect on rainfall is absent) can produce misleading results. This interpretation is supported by failure of predicting the seasonal distributions for Bergen (Fig. 9.9b).

These results may nevertheless give an indication of changes that can be expected in the distribution functions for a number of locations for which the local orography does not play a special role. A future warming and a trend toward a wetter climate can also lead to more heavy precipitation event (Fig. 9.10).

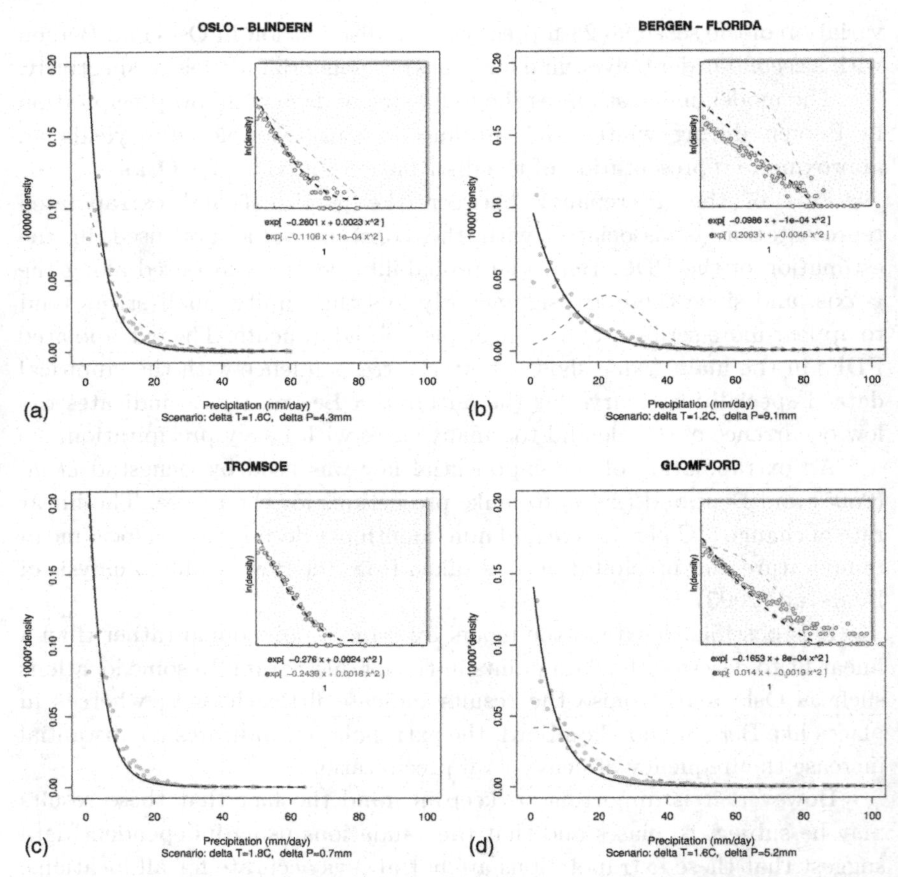

Fig. 9.10. Downscaled PDFs for present-day and future climate for four sites in Norway.

9.7 Further Reading

Stephenson *et al.* (1998) gives a discussion on the distribution of the Indian rainfall data. Pryor *et al.* (2006) discusses ESD for wind speed distributions and Hayhoe *et al.* (2004) downscaled PDFs for the temperature in California.

Benestad *et al.* (2005) attempted to downscale the scale and shape parameters for the gamma distribution describing the daily rainfall amount in Norway, but the relationship between the large-scale SLP and these parameters was weak. They also looked at the latter type, but assumed an exponential distribution rather than gamma.

Benestad (2007) argued that a linear fit can be successfully used as an approximate description here, although a gamma fit may often yield a more accurate representation of the upper tails of the precipitation distribution. Nevertheless, the exponential law is not very different to the more commonly used gamma distribution when its parameters are fitted to provide a best-fit to the daily precipitation amounts (Fig. 9.8).

A readable introduction to probability is given in Chapters 1 and 2 by Wilks (1992, 1995), and in Chapter 4 various commonly used theoretical distribution functions are discussed. The topics on probability and distributions are also covered in von Storch and Ziwers (1999, Secs. I.2 and I.3). A useful reference is also the *Numerical Recipes* (Press *et al.*, 1989, Chapter 13, pp. 548–553). The Gumbel distribution is also discussed on pp. 45–50 in von Storch and Zwiers (1999). A good reference on extreme value theory by Stuart Coles can be found on URL: http://www.maths.lancs.ac.uk/c̃oless/. von Storch and Ziwers (1999, pp. 45–50), Wilks (1995, pp. 93–98), and Press *et al.* (1989, pp. 548–553, 590–598).

9.8 Examples

9.8.1 *iid-test*

```
> library(iid.test)
> library(clim.pact)
> data(oslo.dm)
> i1 <- is.element(oslo.dm$dd,1) & is.element(oslo.dm$mm,7)
> i2 <- is.element(oslo.dm$dd,6) & is.element(oslo.dm$mm,7)
> i3 <- is.element(oslo.dm$dd,11) & is.element(oslo.dm$mm,7)
> i4 <- is.element(oslo.dm$dd,16) & is.element(oslo.dm$mm,7)
> i5 <- is.element(oslo.dm$dd,21) & is.element(oslo.dm$mm,7)
> i6 <- is.element(oslo.dm$dd,26) & is.element(oslo.dm$mm,7)
> i7 <- is.element(oslo.dm$dd,31) & is.element(oslo.dm$mm,7)
> Y <-cbind(oslo.dm$precip[i1],oslo.dm$precip[i2],
          oslo.dm$precip[i3],oslo.dm$precip[i4],
          oslo.dm$precip[i5],oslo.dm$precip[i6],oslo.dm$precip[i7])
> iid.test(Y)
```

9.8.2 *Downscaling PDFs for normal distribution*

```
> library(clim.pact)
> x.rng <- c(-50,40)
> y.rng <- c(45,70)
> load("ERA40_t2m_mon.Rdata")
```

```
> GCM <- retrieve.nc("pcmdi.ipcc4.mpi_echam5.sresa1b.run1.monthly.tas_A1.nc",
  v.nam="tas",x.rng=x.rng,y.rng=y.rng)
> attr(GCM$tim,"unit") <- "month"
> T2m <- catFields(t2m,GCM,mon=1)
> eof <- EOF(T2m,lon=x.rng,lat=y.rng)
> m <- KDVH4DS()
> s <- KDVH4DS(method="sd")
> ds.m <- DS(m,eof)
> ds.s <- DS(s,eof)
> M <- mean(m$val,na.rm=TRUE) + seq(0,10,length=100)* ds.m$rate.ds
> S <- mean(s$val,na.rm=TRUE) + seq(0,10,length=100)* ds.s$rate.ds
> plot(c(2000,2100),range(c(M-2*S,M+2*S)),type="n",
    xlab="Year",ylab="PDF")
> grid()
> for (i in 1:length(M)) {
> lines(rep(2000+i,2),M[i]+2.0*c(-S[i],S[i]),lwd=7,col="grey90")
> lines(rep(2000+i,2),M[i]+1.5*c(-S[i],S[i]),lwd=7,col="grey70")
> lines(rep(2000+i,2),M[i]+1.0*c(-S[i],S[i]),lwd=7,col="grey50")
> lines(rep(2000+i,2),M[i]+0.6*c(-S[i],S[i]),lwd=7,col="grey30")
> lines(rep(2000+i,2),M[i]+0.3*c(-S[i],S[i]),lwd=7,col="grey10")
> }
>dev.copy2eps(file="esd_ds-gaus.eps")
```

The above example shows a simple downscaling exercise for the mean and standard deviations of daily January temperature (Fig. 9.6).

9.8.3 *Downscaling PDFs exponential distribution*

The example below and Fig. 9.11 show how a downscaling of PDFs may be used together with a time series to scale the values according to a predicted distribution. Here, the historical observations have been scaled for simplicity, but the idea is to first apply, e.g. an analog model, then downscale the PDF, and finally use a local quantile transformation to scale the values according to the predicted distribution.

```
> # Example showing a simple exercise where historic precip is
> # rescaled to have a new pdf
> data(exp.law1)
> data(oslo.dm)
> a<-DSpdf.exp(oslo.dm,dT=3,dP=1)
> F1<- list(x=a$x,P=a$Fx.obs)
> F2<- list(x=a$x,P=a$Fx.chg)
> y<-CDFtransfer(Y=oslo.dm$precip,CDF.2=F2,CDF.1=F1,plot=TRUE)
> plot(oslo.dm$precip,main=oslo.dm$location,
       y,xlab="Present 24-hr precip",ylab="future 24-hr precip")
> lines(c(0,100),c(0,100),col="grey",lty=2)
> grid()
```

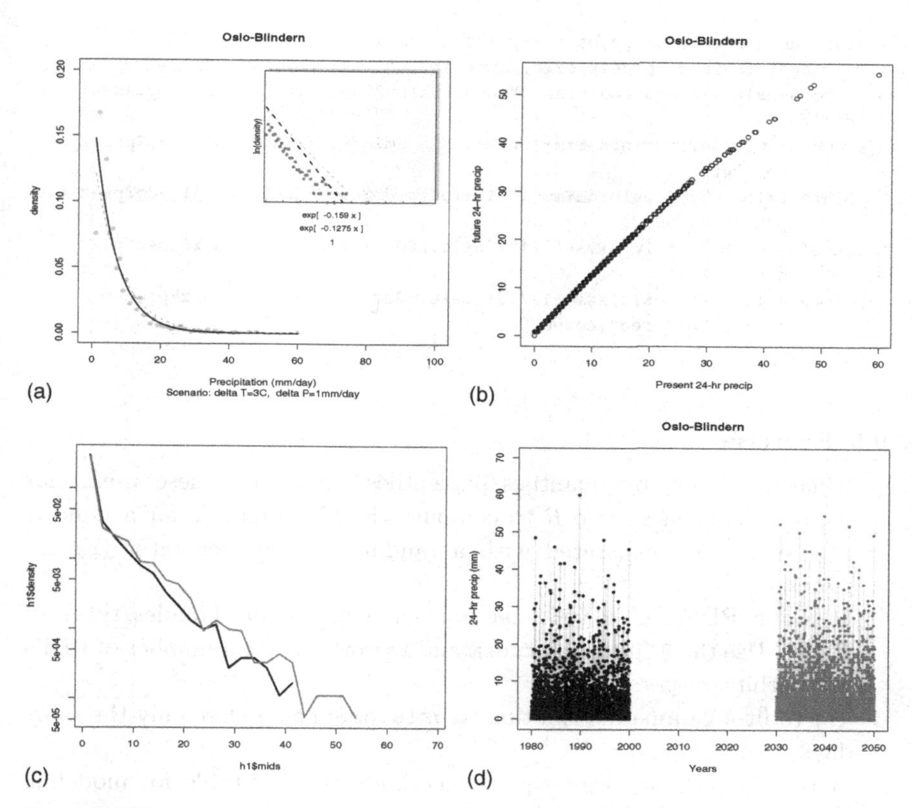

Fig. 9.11. Results from the example demonstrating a simple exercise where historic precip is rescaled to have a new PDF. (a) Predicted PDFs for present climate and the future, (b) local qunatile transform function, (c) original distribution (histogram; black) and after local quantile transform, (d) plot of the time series corresponding to the histograms in (c).

```
> oslo.x <- oslo.dm
> oslo.x$precip <- y
> oslo.x$yy <- oslo.x$yy - min(oslo.x$yy) + 2030
>
> x11()
> breaks <- seq(0,70,by=2.5)
> h1 <- hist(oslo.dm$precip,breaks=breaks)
> h2 <- hist(oslo.x$precip,breaks=breaks)
> plot(h1$mids,h1$density,type="l",lwd=3,log="y")
> grid()
> lines(h2$mids,h2$density,col="red",lwd=3)
>
> x11()
```

```
> plot(range(c(oslo.dm$yy,oslo.x$yy+1),na.rm=TRUE),
>    range(c(oslo.dm$precip,oslo.x$precip),na.rm=TRUE),type="n",
>    main=oslo.dm$location,xlab="Years",ylab="24-hr precip (mm)",ylim=c(0,70))
> grid()
> lines(oslo.dm$yy+(oslo.dm$mm-1)/12+(oslo.dm$dd-1)/365.25,oslo.dm$precip,
>        col="grey")
> points(oslo.dm$yy+(oslo.dm$mm-1)/12+(oslo.dm$dd-1)/365.25,oslo.dm$precip,
>        pch=19,cex=0.6)
> lines(oslo.x$yy+(oslo.x$mm-1)/12+(oslo.x$dd-1)/365.25,oslo.x$precip,
>       col="pink")
> points(oslo.x$yy+(oslo.x$mm-1)/12+(oslo.x$dd-1)/365.25,oslo.x$precip,
>        pch=19,col="red",cex=0.6)
```

9.9 Exercises

1. What is meant by quantiles/percentiles? Why are these useful for studying extremes? Use *R* to compute the 95% quantile for a number of distributions generated with a random number generator (`rnorm`, `rgamma`).
2. Describe PDFs. Can you mention some important families/types of PDFs? Use the `R` functions `rnorm` and `rgamma` to plot a number of PDFs with arbitrary parameters.
3. Try to fit a gamma distribution to `oslo.dm$precip`. Use only the rainy days.
4. Why may ordinary least-squares methods be unsuitable for modeling upper quantiles?
5. Write a code that computes the 90% quantile for each month. Use these values instead of monthly mean, and apply an ESD using `DS` and `DNMI.slp` (tips: make the station series look like a "monthly station" object by taking `oslo.t2m` and replacing the values with the quantiles and changing the element type).
6. Use an analog model to make local scenarios for daily rainfall for the future, and linear models with monthly temperature and precipitation data to estimate the change in the annual mean temperature and precipitation.

Chapter 10

WEATHER GENERATOR

The previous chapters mainly dealt with ESD techniques for monthly climatic variables. For many impact applications and decision support systems daily weather data are required. In particular, extreme events are much more important than the mean climate in the context of climate change and its impact. Daily climatic variables are often required to define extreme climate events (e.g., Moberg *et al.*, 2006).

One of the problems in downscaling daily variables including extremes is due to the fact that daily scale variations are usually too fine scale to be treated by existing global climate model (GCM) and even regional climate model (RCM), which makes statistical downscaling an attractive and interesting alternative. This was clearly demonstrated by Kyselý (2002b) who compared the skills of two GCMs in reproducing extreme high and low temperatures that from statistical downscaling.

One way to downscale daily meteorological variables is to use a weather generator (WG) (e.g., Wilks and Wilby, 1999; Wetterhall *et al.*, 2006). This chapter focuses on modeling and downscaling of daily climatic variables with help of a stochastic model, which will form the basis for the downscaling of extreme covered by next chapter.

Development of WG was started in the early 1960s (e.g., Bailey, 1964). At that time, the researchers were limited to precipitation simulation and the application was mainly found in hydrology (e.g., Gabriel and Neumann, 1962). Today, its application reaches almost every field in assessment of climate impact in conjunction with other models such as agriculture, soil erosion, land use, and ecological systems.

Current models allow simulation of several variables, including precipitation (occurrence and intensity), temperature (maximum, minimum, dew point, and average), radiation, relative humidity, and wind (speed

Fig. 10.1. Weather generators use random numbers in a similar fashion to throwing dice.

and direction). Because the other weather variables such as temperature, radiation, and etc are dependent on precipitation occurrence, the stochastic model for precipitation needs to be specified first.

In this chapter, simulation of precipitation using the Richardson type weather generator is described to illustrate the principles of WG. How WG is used for spatial and temporal downscaling will then be introduced.

10.1 Weather Generator

A weather generator is a stochastic model that can be used to simulate daily weather based on parameters determined by history records (Wilks and Wilby, 1999). Precipitation, maximum temperature, minimum temperature, radiation, and more can all be simulated by WG.

Hydrology has a long tradition in using WG to generate necessary meteorological inputs to hydrological models (e.g., Wight and Hanson, 1991; Clark *et al.*, 2004). Agricultural models (e.g., Wallis and Griffiths, 1997) and ecological models (e.g., Racsko *et al.*, 1991) all make extensive use of WGs.

Recently, WG has gained renewed attention and has been extensively applied to downscale local climate change scenarios (e.g., Semenov, 2007). Such a scenario is useful in studying impacts of climate change on a variety of systems including ecosystem and risk assessment, since it can generate local daily records in the future based on a GCM climate change scenario.

As will be described in the next chapter, WG is an important tool to study extreme climate events and risk analysis. Long-time series are required to study extreme climate events that are rare. However, the length of observed time series is often insufficient to allow reliable estimate of the probability of extreme events.

A WG has the advantage to be able to statistically simulate weather over an extensive period based on using parameters determined from the relatively short history records.

Another application of WG is to simulate meteorological conditions for unobserved locations, which is useful for simulating historical and future climates. This is due to the fact that parameters of a WG are much more stable than the weather itself with regard to locations (Semenov and Brooks, 1999). Thus, the parameters for an unsampled location may be readily interpolated from the surrounding sampled sites. This way a quantitative assessment of impact of climate condition at unsampled location can be realized.

In addition, WG can also be used to do timely assessment of climate variation in terms of a series of daily meteorological variables based on seasonal forecasting of climate.

10.2 Richardson Type WG for Precipitation

Precipitation occurrence and intensity are two processes of a precipitation event (Richardson, 1981). At first, the simulated day is either a wet day or a dry day must be determined. There are three methods to simulate it. Which include two-state (precipitation either occurs or it does not) first-order Markov Chain, multi-order Markov Chain, and Spell length (Srikanthan and McMahon, 2001). Because the first-order Markov Chain is relatively simple and widely applicable, it was used in most of WG including the Richardson type WG.

After the occurrence of precipitation is determined based on transition probabilities, precipitation intensity is simulated. If it is a dry day, precipitation intensity is zero; else precipitation intensity will be simulated based on a stochastic model based on empirical distribution precipitation intensity.

Two-parameter Gamma distribution is usually used for simulation of daily precipitation intensity in the Richardson type WG (e.g., Liao *et al.*, 2004).

10.3 Downscaling Based on WG

Several approaches have been described to use a WG to construct local climate change scenarios (e.g., Wilks, 1992; Semenov and Barrow, 1997). In principle, parameters of the generator need to be modified according to the climate change scenario from a GCM to generate local climate under changed climate conditions.

As an example, the WG used is first calibrated using "average" weather data for a particular region, roughly corresponding to the size of an appropriate GCM grid box, with the resulting parameters describing the statistical characteristics of that region's weather. This "average" weather is calculated using a number of stations from within the relevant region. The WG is also calibrated at each of these individual stations.

The relationship between the parameters of the region and individual stations are established. Second, daily GCM data for the grid box corresponding to the area-average weather data are used to obtain corresponding parameters. The relationship established using observations is used to estimate parameters at individual stations, which allows the generation of scenarios for each station within the area (e.g., Wilks and Wilby, 1999).

Another way to downscale local scenarios is based on a combination of spatial regression downscaling with the use of a local WG (Semenov and Barrow, 1997). Regression downscaling can translate the coarse resolution GCM grid-box predictions of climate change to site-specific values. These values were then used to perturb the parameters of the stochastic WG in order to simulate site-specific daily weather data. This approach produces changes in the mean and variability of climate in a consistent way.

As an example, a WG called ClimGen (e.g., Tubiello *et al.*, 2000) has been used, together with a global monthly temperature data set, to generate regional (0.5*0.5 degree latitude and longitude) scale daily climate in two steps: (1) scaling of the large-scale GCM scenarios on the monthly time scale to the regional scale by using the global monthly data set and the simulated past climate from GCM; (2) decomposing monthly regional climate to daily climate by using ClimGen (Arnell and Osborn, 2006).

Once the PDF is obtained for a variable that represents the future, it is possible to use these in stochastic WGs to produce time series. This type of simulation is also known as "Monte-Carlo simulations," and use a random number generator to random numbers conditioned by the PDF.

It is also important that the WGs produce realistic time structures, such as auro-correlations, as well as realistic wet-spell durations and dry intervals. The probability of a rainy day after a dry day, and vice versa, is important to model correctly.

WGs applied to single locations at the time may not ensure spatial consistency, which can be a problem. One way to solve this may be to apply a stochastic analog model (Zorita *et al.*, 1995) (see Chap. 5), thus prescribing simultaneous values globally.

Soltani and Hoogenboom (2003) evaluated the impact of the length of input data on the quality of the output from the WGs. They found that the length was sensitive to the type of weather generator (WGEN or SIMMETEO) and the parameter, and recommended that at least 15 years of data is required.

10.4 Examples

10.4.1 *Coin-flipping*

Heads (0) and tails (1). For 100 trials, what is the probability that there will be less than four heads?

$$p = 0.5; \quad N = 100;$$

$$\begin{aligned} &[1 + N + N(N-1)/2 + N(N-1)(N-2)/3]0.5^N \\ &= 328451 \times 0.5^{100} = 2.6 \times 10^{-25}. \end{aligned}$$

The chance of getting less than four heads is practically zero: if the remaining life time of our solar system is 10 billion years ($t = 3.15 \times 10^{17}\,$s) and it takes 1 s to sample each set of 100 throws, the chance of obtaining four or less heads is $P_r(X \leq 4) = 8 \times 10^{-8}$.

10.5 Exercises

1. Use the `rnorm` and `rgamma` to generate series of synthetic data for temperature and precipitation, respectively, by taking the mean and

standard deviations from an observed daily temperature series and shape and scale for precipitation (wet days only).

2. Why are only wet days used for the precipitation fit?
3. Use the data from the exercise above and introduce nonrainy days according to the observed frequency.
4. How do the simulated time structure compare with the observed, in terms of length of dry spells/wet spells and transition probabilities?

Chapter 11

IMPLEMENTING ESD

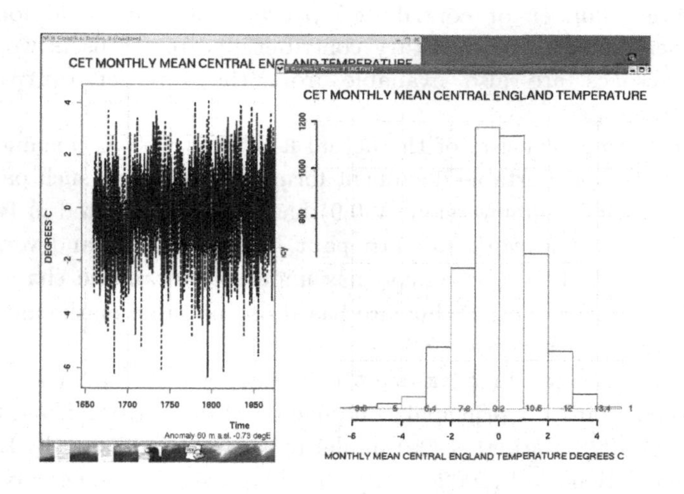

11.1 Numerical Tools

11.1.1 *R*

11.1.1.1 *clim.pact*

The tool, `clim.pact`,[1] tailored for climate data has been written in order to facilitate more efficient, easier, and faster analysis. This tool consists of an R-package (Ellner, 2001; Gentleman and Ihaka, 2000), and provides additional functions in the R environment for climate analysis, visualization,

[1]Available from the CRAN Internet site (http://cran.r-project.org/) under the link to "contributed packages."

and empirical downscaling. `R` is a statistical software freely available from the Internet (`http://www.R-project.org/`), sometimes referred to as the "a GNU-version of Splus."

Documentation on `R` is also freely available from the Internet,[2] but is also installed locally as (hypertext mark-up language) HTML pages, so that a Web-browser can be used in an online-help facility (activated through the `R` commando: "help.start()").

The user-support in `R` is impressive, especially considering it being a freely available software with no license restrictions. An advantage of `R`-packages is that the `R`-environment combines their functionality with the large number of functions available in the `R`-environment as well as from other packages.

A large number of contributed packages are available for the `R` environment, which are voluntary contributions from `R`-users worldwide. These packages are also available from the Internet (`http://www.R-project.org/`).

An `R`-package consists of the actual `R`-code as well as documentation in the HTML and portable document format (PDF). One such package is `clim.pact`, whose pilot version (V.0.9) has been documented in Benestad (2004b, 2003c). The work on `clim.pact` has progressed, and version 2.2 is now released. The new version has undergone dramatic changes with respect to the user-syntax,[3] but are based on the same mathematical and logical framework as version 0.9.

The downscaling in `clim.pact` can easily be set up in a way that incorporates common principal components (Flury, 1988; Sengupta and Boyle, 1993, 1998; Barnett, 1999) similar to the work published by Benestad *et al.* (2002); Benestad (2002a,b, 2001b). The main difference between the method used to derive these published results and `clim.pact` is that the default method used by `clim.pact` is linear regression instead of CCA[4]-based models.

Furthermore, `clim.pact` utilizes a stepwise screening procedure that aims to minimize the Akaike information criterion (AIC) (Wilks, 1995, pp. 300–302), whereas the CCA-based results carried out a stepwise screening based on the correlation coefficients in a cross-validation analysis (Wilks, 1995, pp. 194–198).

[2] "An Introduction to `R`," "the `R` language definition," "Writing `R` Extensions," "`R` Data Import/Export," "`R` Installation and Administration," "The `R` Reference Index," in addition to frequently asked questions (FAQs), Contributed, and newsletters.
[3] How the functions are called.
[4] CCA = canonical correlation analysis.

This package also incorporates polynomial descriptions of climatic trends (Benestad, 2003e) and provides an easy way to pre-process climate data (spatial maps and station records), as well as "house keeping" in terms of matching time stamps, etc. Additional features include composites and spatial correlation analysis.

The purpose of this book is primarily to give the readers an up-to-date documentation on the use of `clim.pact` (Benestad, 2003b, 2004b).

11.1.1.2 *Rclim*

Another useful `R`-package is `Rclim`, available from http://www.met.reading.ac.uk/cag/rclim/. This package is less geared toward ESD, but more focused on statistical analysis.

11.1.2 *SDSM*

The SDSM[5] ESD tool is constructed for Windows platforms by Rob Wilby and Chris Dawson. SDSM has been used extensively for impact studies and is "a user-friendly software package designed to implement statistical downscaling methods to produce high-resolution monthly climate information from coarse-resolution climate model (GCM) simulations. The software also uses weather generator methods to produce multiple realizations (ensembles) of synthetic daily weather sequences." Recent study based on SDSM includes that of Wilby and Harris (2006).

11.1.3 *ClimateExplorer*

The climateexplorer tool (http://climexp.knmi.nl/) developed by Geert Jan van Oldenborgh at the KNMI offers a nice suite of analyses and a large database for statistical studies. Although this is not tailored for ESD specifically, it can nevertheless provide useful statistics and diagnostics.

11.1.4 *ENSEMBLE-S2D*

The EU-project ENSEMBLES has established an Internet portal for ESD on the URL www.meteo.unican.es/ensembles. This internet page has a facility, which allows ESD carried out online.

[5] http://www-staff.lboro.ac.uk/~cocwd/SDSM/ManualSDSM.pdf

11.2 Gridding the Results

11.2.1 *Kriging*

Benestad (2002b) used kriging analysis (Matheron, 1963) in order to construct spatial maps describing how future warming (i.e. the multimodel ensemble mean) may vary geographically. Kriging is a standard method used for spatial interpolation in geosciences, and an evaluation of the kriging methodology is outside the scope of this book.

Although this kind of spatial interpolation gives an approximately realistic representation, it does not take into account the fact that local warming rates may vary with the distance from the coast, altitude, latitude, and longitude. In order to produce more realistic maps, this geographical information must be included in the spatial analysis.

Recently Chen *et al.* (2007) used kriging to create a gridded baseline climatology of monthly mean temperature for Sweden. The same method can also be used to create a gridded future scenario given the downscaled future temperatures at the stations.

11.2.2 *Residual kriging*

The kriging analysis by Matheron (1963) can be applied to the residuals of the regression analysis in order to spatially interpolate the part of the trends that could not be related to the geographical parameters (using the `geoR`-package for `R`). Hence, the maps represented more than just the geographical variance accounted for by the multiple regressions used for geographical modeling, since interpolated residuals are added to the prediction to recover most of the signals.

The results presented by Benestad (2004e) involved further "refinement" to those by Benestad (2002b) (Fig. 11.1) by taking into consideration geographical information not utilized in the earlier work, in addition to examining conditional probability estimates instead of multimodel ensemble mean values. The results derived for each site was used in a geographical model based on a multiple regression analysis against distance from the coast, altitude, latitude, and longitude. The geographical regression models will henceforth be referred to as "GRMs." The details of the model calibration performed by Benestad (2004e) is reproduced in Table 11.1, and are listed in the form of coefficient estimates, standard error estimates, t-values and probabilities (p-value) of null-hypothesis (zero coefficient) being true.

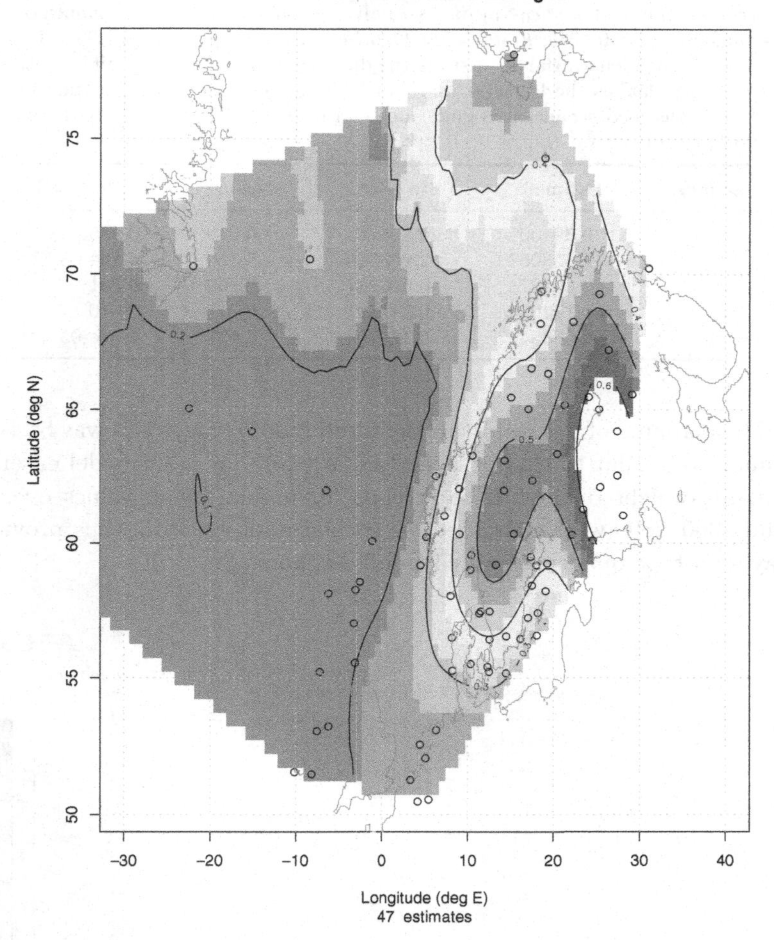

Fig. 11.1. Example of a kriging analysis for mapping the results (Benestad, 2002b).

Also shown are the estimates for the R^2 (variance explained), the F-statistic ("strength" of the regression), the degrees of freedom, and the p-value for the entire regression. It is important to keep in mind that the geographical models derived here may not be valid for other parts of the world. A kriging analysis similar to that of Benestad (2002b) was used for the spatial interpolation of the residuals from the regression analysis.

Table 11.1. The analysis of variance (ANOVA) of the geographical multiple regression model, $y = c_0 + c_1$ dist + c_2 alt + c_3 lat + c_4 lon, for the month of January. $R^2 = 0.4239$; F-stat = 20.23; on 4 and 110 DF; p-value = 1.635e-12. The independent variable "dist" is the distance from the coast, "alt" is the altitude, "lat" is the latitude, and "lon" is the longitude of the location of the downscaled scenarios. Significance codes: 0 "***" 0.001 "**" 0.01 "*" 0.05 "." 0.1 " " 1.

January	Estimate	Std. Error	t-value	Pr(> |t|)
c_0	0.394001	20.805723	0.019	0.985
c_1	17.358068	2.998599	5.789	6.80e-08 ***
c_2	−0.005374	0.006237	−0.862	0.391
c_3	0.174924	0.342191	0.511	0.610
c_4	0.583544	0.126104	4.627	1.02e-05 ***

The mapping of the results in Benestad (2005) (Fig. 11.2) was based on the analysis similar to that of Benestad (2004e), but multimodel ensemble w_i quality-weighted mean linear trends for annual mean values over the period 2000–2099 were estimated for each station location, thus providing a Bayesian-type quality-weighted trend estimate.

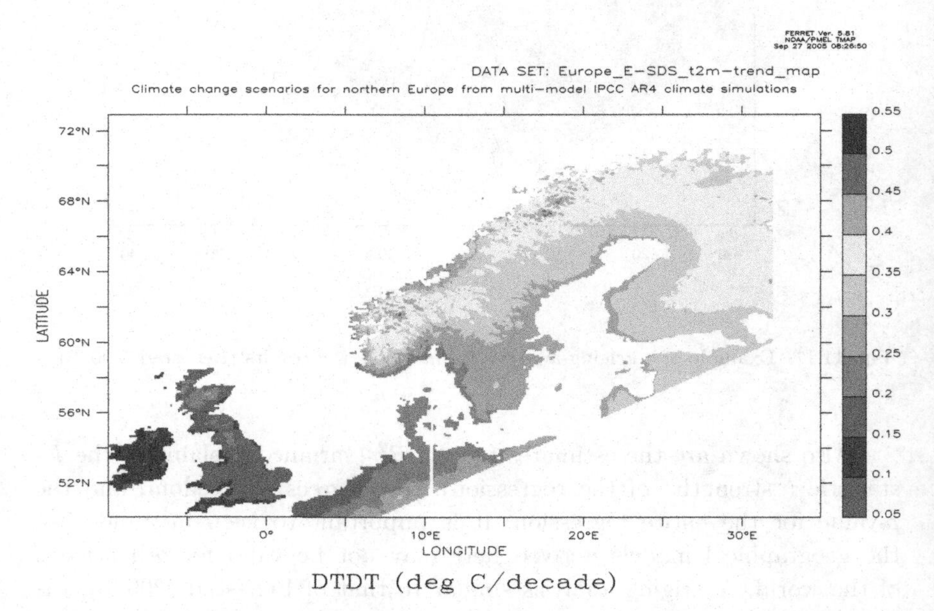

Fig. 11.2. Example of GRM results plus residual kriging analysis from Benestad (2005). Combining GRM with residual kriging yields more detailed maps than plain kriging.

The technique used in Benestad (2005) differed further from that of Benestad (2004e) by using the square-root distance from the coast ($\sqrt{d}$) as opposed to a linear relation with distance, and including two additional geographical predictors: north–south slope and east–west slope.

Whereas Benestad (2004e) used longitude and latitude as two independent variables representing the coordinates, Benestad (2005, 2007) used eastings and northings: east–west and north–south displacements from the central point of the set of locations, in units of 10 km. Due to the Earth's curvature, a difference of 1° at high latitudes corresponds to a smaller zonal displacement than that at lower latitudes.

The east–west and north–south slopes were estimated through a stepwise multiple regression fit to $N_\theta = N_\phi = 35$ harmonics to the topographical cross-sectional profile following Eq. (11.2) and then solving for the derivatives according to Eq. (11.3):

$$\begin{aligned} z(\theta) &= z_0 + \sum_{i=1}^{N_\theta} [a_\theta(i)\cos(\omega_\theta(i)\theta) + b_\theta(i)\sin(\omega_\theta(i)\theta)], \\ z(\phi) &= z_0 + \sum_{i=1}^{N_\phi} [a_\phi(i)\cos(\omega_\phi(i)\phi) + b_\phi(i)\sin(\omega_\phi(i)\phi)], \end{aligned} \tag{11.1}$$

$$\begin{aligned} \frac{\partial \hat{z}(\theta)}{\partial \theta} &= \sum_{i=1}^{N_\theta} \omega_\theta(i)[-\hat{a}_\theta(i)\sin(\omega_\theta(i)\theta) + \hat{b}_\theta(i)\cos(\omega_\theta(i)\theta)], \\ \frac{\partial \hat{z}(\phi)}{\partial \phi} &= \sum_{i=1}^{N_\phi} \omega_\phi(i)[-\hat{a}_\phi(i)\sin(\omega_\phi(i)\phi) + \hat{b}_\phi(i)\cos(\omega_\phi(i)\phi)]. \end{aligned} \tag{11.2}$$

Since spherical coordinates were used, a transformation was done to x- and y-coordinates following Eq. (11.3):

$$\begin{aligned} \frac{d\hat{p}(x)}{dx} &= \frac{1}{a\cos(\phi)}\frac{d\hat{p}(\theta)}{d\theta}, \\ \frac{d\hat{p}(y)}{dy} &= \frac{1}{a}\frac{d\hat{p}(\phi)}{d\phi}. \end{aligned} \tag{11.3}$$

The harmonics fit, differentiation, and transformation were done in the `R`-environment, using the `geoGrad` function in the contributed `cyclones`-package (version 1.1-4).

The GRMs can be assessed further in split-sample tests, where part of the data was used for model calibrating (dependent) and the rest as independent data for evaluation (Benestad, 2004a). Figure 11.3 shows a

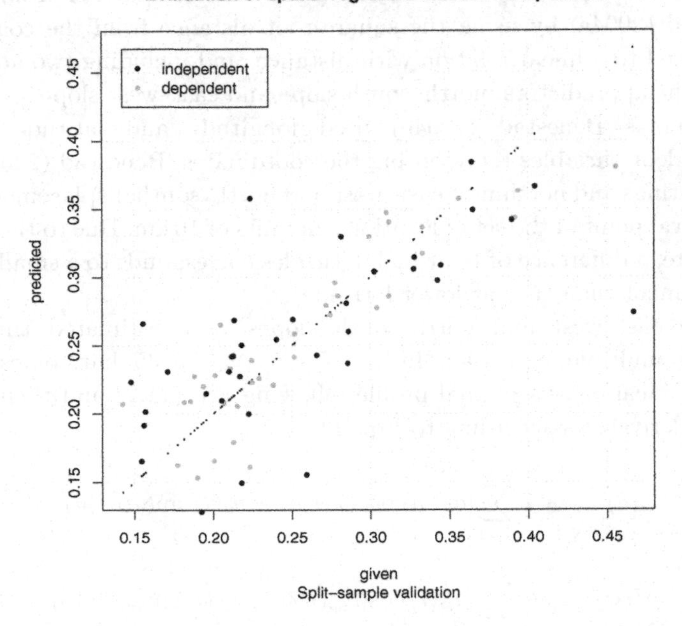

Fig. 11.3. Example of GRM results plus residual kriging analysis (Benestad, 2005).

scatter plot between the original data and the predicted values for both the dependent (grey) and independent data (black), and provides a verification of the GRM.

A kriging analysis can be applied to the residuals of the GRM in order to spatially interpolate the part of the trends that could not be related to geographical parameters (using the `geoR`-package from CRAN).

11.2.3 *GIS-packages*

There are commercial as well as freely available open-source geostatistical applications for Geographical Information System (GIS) analysis. There is a package called GRASS (http://grass.itc.it/statsgrass/index.html) and an `R`-package with the same name, that works as an interface between the application and `R`.

There are also several `R`-packages for implementing kriging, such as `geoR`, `geoRglm`, `GeoXp`, `gstat sgeostat`, and `spgrass6`. Basically, as long as there are results for a number of locations, it is possible to use GIS-type tools to produce maps, and the larger the number of locations, the more robust the results would be.

ACKNOWLEDGEMENTS

Benestad and Hanssen-Bauer would like to acknowledge the support from the Norwegian Meteorological Institute (NMI) and the Norwegian Research Council (NRC), as a large part of the material presented in the book is based on work associated with both NRC research projects and the NMI.

Deliang Chen wishes to thank the Swedish Research Council, the Swedish Foundation for International Cooperation in Research and Higher Education, the Swedish International Development Cooperation Agency, the Swedish Rescue Services Agency, the Swedish Energy Agency, Chinese Academy of Sciences, Beijing Climate Center, China Meteorological Administration and the Faculty of Science at the University of Gothenburg through grants (e.g. Tellus) related to the downscaling works.

APPENDIX

AMAP Demonstration

Listing of R-code used for demonstration at the AMAP workshop in Oslo, May 14–16, 2007 (the prompt is not shown here):

```
rm(list=ls()) # clear the memory

gcm.t2m.1<-"pcmdi.ipcc4.ncar_ccsm3_0.sresa1b.run1.monthly.tas_A1.nc"
gcm.t2m.2<-"pcmdi.ipcc4.mpi_echam5.20c3m.run1.monthly.tas_A1.nc"
gcm.t2m.3<-"pcmdi.ipcc4.mpi_echam5.20c3m.run1.daily.
            tas_A2_1961-1980.nc"
gcm.t2m.4<-"pcmdi.ipcc4.mpi_echam5.sresa1b.run2.daily.
            tas_A2_2046-2065.nc"
gcm.slp.1<-"pcmdi.ipcc4.mpi_echam5.20c3m.run1.daily.
            psl_A2_1961-1980.nc"
gcm.slp.2<-"pcmdi.ipcc4.mpi_echam5.sresa1b.run2.daily.
            psl_A2_2046-2065.nc"
era40.t2m<-"era40_t2m.nc"
era40.slp<-"era40_slp.nc"

# To get the on-line help
help.start()

# Read the CET from the Internet:
cet <- getHadObs()
plotStation(cet)
data(DNMI.slp)

?corField
corField(DNMI.slp,cet,mon=1)

# Read on-line station data from the NARP project
NARP <- getnarp()
print(NARP$name)
```

```
nuuk <- getnarp("Nuuk")
plotStation(nuuk)

# Construct a 'station object':
? station.obj

rnd <- station.obj(x=matrix(rnorm(100*12),100,12),yy=1901:2000,
                   obs.name="Random",unit="dimensionless",ele=101,
                   mm=NULL,station=NULL,lat=60,lon=0,alt=NULL,
                   location="unspecified",wmo.no=NULL,
                   start=NULL,yy0=NULL,country=NULL,ref=NULL)
plotStation(rnd)
corField(DNMI.slp,rnd,mon=1)

# Retrieve SSTs for the North Atlantic.
print("")
data(DNMI.sst)
corField(DNMI.sst,nuuk,mon=c(12,1,2))

# Field object handling
data(eof.slp)
plotEOF(eof.slp)
SLP <- EOF2field(eof.slp)
corField(SLP,nuuk,mon=1)

# Retrieve and handle any field object:
GCM.t2m.1<-retrieve.nc(gcm.t2m.1,v.nam="tas",x.rng=c(0,40),
           y.rng=c(60,90))
GCM.t2m.1$yy <- GCM.t2m.1$yy + 2000
mapField(GCM.t2m.1)
mT2m <- meanField(GCM.t2m.1)
map(mT2m)

#-------------------------------------------------------------
# ESD for Monthly data:
#-------------------------------------------------------------

# simple ESD:
data(DNMI.t2m)
eof.t2m.1 <- EOF(DNMI.t2m,mon=1)
plotEOF(eof.t2m.1)
ds.nuuk.1 <- DS(nuuk,eof.t2m.1)

# simple ESD with random values:
ds.rnd <- DS(rnd,eof.t2m.1)
```

```
# combinde fields & perform ESD:
T2m.1 <- catFields(DNMI.t2m,GCM.t2m.1)
eof.t2m <- EOF(T2m.1,mon=1)
plotEOF(eof.t2m)
ds.nuuk <- DS(nuuk,eof.t2m)

# read daily data for ERA40
print("")
ERA40.t2m.day<-retrieve.nc(era40.t2m,v.nam="p2t",x.rng=c(0,40),
               y.rng=c(60,90))
ERA40.t2m <- monthly(ERA40.t2m.day)
eof.t2m.2 <- EOF(ERA40.t2m,mon=1)
plotEOF(eof.t2m.2)

region.t2m <- catFields(DNMI.t2m,lon=c(0,40),lat=c(60,90))
eof.t2m.3 <- EOF(region.t2m,mon=1)

hopen <- getnarp("Hopen")
plotStation(hopen,what="t")
corField(ERA40.t2m,hopen,mon=1)
corField(region.t2m,hopen,mon=1)

ds.hopen.1 <- DS(hopen,eof.t2m.2)
ds.hopen.2 <- DS(hopen,eof.t2m.3)

T2m.eragcm <- catFields(ERA40.t2m,GCM.t2m.1)
T2m.dnmigcm <- catFields(region.t2m,GCM.t2m.1)
eof.T2m.1 <- EOF(T2m.eragcm,mon=1)
eof.T2m.2 <- EOF(T2m.dnmigcm,mon=1)

ds.Hopen.1 <- DS(hopen,eof.T2m.1)
ds.Hopen.2 <- DS(hopen,eof.T2m.2)

GCM.t2m.2<-retrieve.nc(gcm.t2m.2,v.nam="tas",x.rng=c(0,40),
           y.rng=c(60,90))
print(summary(GCM.t2m.2))
print(table(GCM.t2m.2$id.t))
GCM.t2m.2$id.t[] <- "ECHAM5"
attr(GCM.t2m.2$tim,"unit") <- "month"
T2m.eragcm <- catFields(ERA40.t2m,GCM.t2m.2)
T2m.dnmigcm <- catFields(region.t2m,GCM.t2m.2)
eof.T2m.3 <- EOF(T2m.eragcm,mon=1)
eof.T2m.4 <- EOF(T2m.dnmigcm,mon=1)

ds.Hopen.3 <- DS(hopen,eof.T2m.3)
```

```
ds.Hopen.4 <- DS(hopen,eof.T2m.4)

hopen.1 <- ds2station(ds.Hopen.1)
hopen.2 <- ds2station(ds.Hopen.2)
hopen.3 <- ds2station(ds.Hopen.3)
hopen.4 <- ds2station(ds.Hopen.4)

plotStation(hopen.3,col="blue",what="t",type="b",pch=20,lty=1,lwd=1,
         trend=FALSE,std.lev=FALSE)
plotStation(hopen.4,col="red",add=TRUE,what="t",type="b",pch=20,
            lty=1,lwd=1,trend=FALSE,std.lev=FALSE)
plotStation(hopen.1,col="lightblue",add=TRUE,what="t",type="b",
            pch=21,lty=1,lwd=1,trend=FALSE,std.lev=FALSE)
plotStation(hopen.2,col="pink",add=TRUE,what="t",type="b",pch=21,
            lty=1,lwd=1,trend=FALSE,std.lev=FALSE)
plotStation(hopen,add=TRUE,what="t",type="b",pch=19,lty=3,lwd=1,
            trend=FALSE,std.lev=FALSE)

#-------------------------------------------------------------------
# ESD for Daily data:
#-------------------------------------------------------------------

data(oslo.dm)
ERA40.t2m.day<-retrieve.nc(era40.t2m,v.nam="p2t",x.rng=c(0,40),
                           y.rng=c(50,70))
GCM.t2m.day.1<-retrieve.nc(gcm.t2m.3,v.nam="tas",x.rng=c(0,40),
                           y.rng=c(50,70))
GCM.t2m.day.2<-retrieve.nc(gcm.t2m.4,v.nam="tas",x.rng=c(0,40),
                           y.rng=c(50,70))
T2m.day <- catFields(GCM.t2m.day.1,GCM.t2m.day.2,demean=FALSE)
T2m.day <- catFields(ERA40.t2m.day,T2m.day)
eof.t2m.day <- EOF(T2m.day,mon=2)
ds.oslo <- DS(oslo.dm,eof.t2m.day)

# Analog model for preciptation
library(anm)
ERA40.slp.day<-retrieve.nc(era40.slp,v.nam="slp",x.rng=c(0,40),
                           y.rng=c(50,70))
GCM.slp.1 <- retrieve.nc(gcm.slp.1,v.nam="psl",x.rng=c(0,40),
                         y.rng=c(50,70))
GCM.slp.2 <- retrieve.nc(gcm.slp.2,v.nam="psl",x.rng=c(0,40),
                         y.rng=c(50,70))
SLP.day <- catFields(GCM.slp.1,GCM.slp.2,demean=FALSE)
SLP.day <- catFields(ERA40.slp.day,SLP.day)
eof.slp.day <- EOF(SLP.day,mon=2)
```

```
anm.method <- "anm.weight"
param <- "precip"

a.lm <- DS(preds=eof.slp.day,dat=oslo.dm,
           plot=FALSE,lsave=FALSE,param=param,
           ldetrnd=FALSE, rmac=FALSE)

i.eofs <- as.numeric(substr(names(a.lm$step.wise$coefficients)[-1],
          2,3))
a.mam <- DS(preds=eof.slp.day,dat=oslo.dm,i.eofs=i.eofs,
              method=anm.method,swsm="none",
              predm="predictAnm",param=param,
              lsave=FALSE,ldetrnd=FALSE,rmac=FALSE)

ictl <- a.mam$yy.gcm <= 2000
dT<-  mean(ds.oslo$pre.gcm[!ictl],na.rm=TRUE) -
      mean(ds.oslo$pre.gcm[ictl],na.rm=TRUE)
dP <- 0.20*mean(oslo.dm$precip,na.rm=TRUE)

# Downscale PDF.
a<-DSpdf.exp(oslo.dm,dT=dT,dP=dP)
F1<- list(x=a$x,P=a$Fx.obs)
F2<- list(x=a$x,P=a$Fx.chg)
y.ctl<-CDFtransfer(Y=a.mam$pre.gcm[ictl],CDF.2=F1,plot=TRUE)
y.sce<-CDFtransfer(Y=a.mam$pre.gcm[!ictl],CDF.2=F2,plot=TRUE)
plot(c(y.ctl,y.sce),pch=19,cex=0.7)
points(y.ctl,pch=19,col="grey",cex=0.7)
```

Some of these lines are run by typing `demo(ESD.demo)`.

General ESD-Script

```
rm(list=ls())
library(clim.pact)

sdate2yymmdd <- function(sdate) {
  yy <- as.numeric(substr(sdate,1,4))
  mm <- as.numeric(substr(sdate,6,7))
  dd <- as.numeric(substr(sdate,9,10))
  yymmdd <- data.frame(year=yy,month=mm,day=dd)
  invisible(yymmdd)
}

ds.one <- function(ele=101,cmons=1:12,silent=FALSE,new.plot=TRUE,
```

```
                         do.20c=TRUE,do.a1b=TRUE,qc=FALSE,
                         scen="sresa1b",post=FALSE,passwd=NULL,
                         predictand = "narp",station="Tasiilaq",
                         test=FALSE,replace=NULL,ipcc.ver="AR4.",
                         off=FALSE,LINPACK=TRUE,downloads.since=NULL,
                         C20="20c3m",
                         dir="data/IPCC_AR4/GCMs/",op.path="output",
                         v.names=c('tas','pr'),force=FALSE) {

if (!do.20c) C20 <- NULL

if ( (op.path!="./") &  !exists(op.path) ) {
  print(paste("Create new directory (1):",op.path))
  dir.create( op.path )
}

v.nam <- switch(as.character(ele),'101'=v.names[1],'601'=v.names[2])
print("Get ERA40")
if (ele==101) {
  load("ERA40_t2m_mon.Rdata")
  era40.t2m <- catFields(t2m,lon=c(-90,50),lat=c(40,75));rm(t2m)
} else if (ele==601) {
  load("ERA40_prec_mon.Rdata")
  era40.t2m <- catFields(prec,lon=c(-90,50),lat=c(40,75));rm(prec)
  era40.t2m$dat <- era40.t2m$dat * 1.599876e-05/0.3675309
}

#print("Get NCEP")
#ncep.t2m <- retrieve.nc("/data/NCEP/air.mon.mean.nc")

gcms <- list.files(path=dir,pattern=v.nam)
print(gcms)
gcms <- gcms[grep(".nc",gcms)]
gcms <- gcms[grep(v.nam,gcms)]
if (length(grep(".nc.part",gcms))>0) gcms<-gcms[-grep(".nc.part",
    gcms)]
if (!is.null(C20)) gcms1<-reverse(gcms[grep(C20,lower.case(gcms))])
                         else gcms1 <- NULL
gcms2 <- reverse(gcms[grep(scen,gcms)])

if (station=="Vardoe") {
  if (!is.null(C20)) gcms1 <- gcms1[-15]
  gcms2 <- gcms2[-10]
}
if (test) {
```

```
  if (!is.null(replace)) {
    print(paste("source(",replace,")"))
    source(replace)
  }

  if (!is.null(C20)) gcms1 <- gcms1[14]
  gcms2 <- gcms2[1]
}

if (!is.null(downloads.since)) {
   print(paste("Only inlcude data downloaded since",downloads.since))
   if (!is.null(C20)) times.gcms1 <-
  substr(as.character(file.info(paste(dir,gcms1,sep="")))$mtime,1,10)
   times.gcms2 <-
  substr(as.character(file.info(paste(dir,gcms2,sep=""))$mtime),1,10)
   print(times.gcms2)
   since.date <- sdate2yymmdd(downloads.since)
   if (!is.null(C20)) {
     i1 <- ((sdate2yymmdd(times.gcms1)$year == since.date$year) &
            (sdate2yymmdd(times.gcms1)$month == since.date$month) &
            (sdate2yymmdd(times.gcms1)$day >= since.date$day) ) | (
            (sdate2yymmdd(times.gcms1)$year == since.date$year) &
            (sdate2yymmdd(times.gcms1)$month > since.date$month) ) |
            (sdate2yymmdd(times.gcms1)$year > since.date$year)
     print(paste(sum(i1),"runs since",since.date$day,
          since.date$month,since.date$year))
   }
   i2 <- ((sdate2yymmdd(times.gcms2)$year == since.date$year) &
          (sdate2yymmdd(times.gcms2)$month == since.date$month) &
          (sdate2yymmdd(times.gcms2)$day >= since.date$day) ) | (
          (sdate2yymmdd(times.gcms2)$year == since.date$year) &
          (sdate2yymmdd(times.gcms2)$month > since.date$month) ) |
          (sdate2yymmdd(times.gcms2)$year > since.date$year)
   if (!is.null(C20)) gcms1 <- gcms1[i1]
   print(paste(sum(i2),"runs since",since.date$day,
         since.date$month,since.date$year))
   gcms2 <- gcms2[i2]
}

if (!is.null(C20)) print(gcms1)
if (!silent) { print("GCMS:"); print(gcms2)}

if (class(station)[1]=="character") {
  print(station)
  if (lower.case(predictand)=="nordklim") {
```

```
    obs <- getnordklim(station,ele=ele)
  } else if (lower.case(predictand)=="nacd") {
    obs <- getnacd(station,ele=ele)
  } else if (lower.case(predictand)=="narp") {
    obs <- getnarp(station,ele=ele)

  }
} else if (class(station)[1]=="station") {
  print(paste("Use given station object",station$location))
  obs <- station
  station <- obs$location
}

print(obs$location)
if (sum(is.na(obs$val))>0) obs$val[length(obs$yy),] <- NA

x.rng <- c(max(c(obs$lon-30,-180)),min(c(obs$lon+15,180)))
y.rng <- c(max(c(obs$lat-15,-90)), min(c(obs$lat+15,90)))
print(x.rng); print(y.rng)

slsh <- instring("/",obs$location)
if (slsh[1] > 0) {
  obs$location <- substr(obs$location,1,slsh[1]-1)
}

fname.png <- paste("ds_one_",ipcc.ver,predictand,strip(obs$location),
                   obs$station,ele,".png",sep="")
print(fname.png)
if (options()$device=="png") bitmap(file=fname.png,type="png256",
                                    res=300) else newFig()

plot(c(1890,2100),c(min(rowMeans(obs$val[,cmons]),na.rm=TRUE)-2,
                    max(rowMeans(obs$val[,cmons]),na.rm=TRUE)+7),
     type="n",main=obs$location,ylab=obs$obs.name,xlab="time")
grid()
obs.ts <- plotStation(obs,what="t",add=TRUE,col="grey20",type="p",
                      pch=19,l.anom=FALSE,mon=cmons,trend=TRUE,std.
                      lev=FALSE)

print(obs$location)
i.gcm <- 0

if (!is.null(C20)) {
  for (gcm in gcms1) {
    i.gcm <- i.gcm + 1
```

```
    print(gcm)
    dot <- instring(".",gcm)
    if (length(dot)>=3) {
      run<- substr(gcm,dot[4]+1,dot[4]+4)
      gcm.nm <- substr(gcm,dot[2]+1,dot[3]-1)
    } else {
        dash <- instring("_",gcm)
        gcm.nm <- substr(gcm,1,dash[1]-1)
        run<- substr(gcm,dash[1]+1,dash[2]-1)
    }

    run<- substr(gcm,dot[4]+1,dot[4]+4)
#print(obs$location)
    slsh <- instring("/",obs$location)
    if (slsh[1] > 0) {
     obs$location <- substr(obs$location,1,slsh[1]-1)
    }
#print(obs$location)
    subdir <- paste(predictand,strip(obs$location),obs$station,ele,
                    sep="")
    fname <- paste(op.path,"/",subdir,"/ds_one_",ipcc.ver,predictand,
                   strip(obs$location),obs$station,ele,".",
                   gcm.nm,"c20.",run,".",sep="")
    if (!file.exists( paste(op.path,"/",subdir,sep="") )) {
      if (!silent) print(paste("Creating  (2)",op.path,"/",subdir,
                               sep=""))
      dir.create( paste(op.path,"/",subdir,sep="") )
    } else if (!silent) print(paste(op.path,"/",subdir," exsists...",
                                    sep=""))
    if (!silent) print(fname)
    if (!file.exists(paste(fname,"txt",sep="")) &
        !file.exists(paste(fname,"Rdata",sep="")) | (force)) {

      GCM <- retrieve.nc(paste(dir,gcm,sep=""),x.rng=x.rng,
                         y.rng=y.rng,v.nam=v.nam,silent=FALSE)
      class(GCM) <- c("field","monthly.field.object")
      attr(GCM$tim,"unit") <- "month"; GCM$dd[] <- 15

      ds.station <- objDS(era40.t2m,GCM,obs,plot=FALSE,
            qualitycontrol=qc,silent=silent,LINPACK=LINPACK)
      x <- ds.station$station

      print(paste("Saving in",fname))
      save(file=paste(fname,"Rdata",sep=""),x,ds.station)
      ds.scen <- data.frame(Year=x$yy,
```

```
                 Jan=round(x$val[,1],2),Feb=round(x$val[,2],2),
                 Mar=round(x$val[,3],2),Apr=round(x$val[,4],2),
                 May=round(x$val[,5],2),Jun=round(x$val[,6],2),
                 Jul=round(x$val[,7],2),Aug=round(x$val[,8],2),
                 Sep=round(x$val[,9],2),Oct=round(x$val[,10],2),
                 Nov=round(x$val[,11],2),Dec=round(x$val[,12],2))
      write.table(ds.scen,file=paste(fname,"txt",sep=""),
                 row.names = FALSE,quote = FALSE, sep="\t ")
    } else {
      load(paste(fname,"Rdata",sep=""))
     print(paste(fname,"exists, reading from file."))
    }
    plotStation(x,what="t",add=TRUE,col="grey40",type="l",lwd=2,
               lty=1,l.anom=FALSE,mon=cmons,trend=FALSE,std.
               lev=FALSE)
    plotStation(obs,what="t",add=TRUE,col="grey20",type="p",pch=19,
               l.anom=FALSE,mon=cmons,trend=TRUE,std.lev=FALSE)
    plotStation(obs,what="t",add=TRUE,col="grey20",type="l",lwd=1,
               lty=3,l.anom=FALSE,mon=cmons,trend=TRUE,std.
               lev=FALSE)
  }
}
print(paste("Scenarios & ipcc version:",ipcc.ver))
i.gcm <- 0
for (gcm in gcms2) {
  i.gcm <- i.gcm + 1
  print(gcm)

  dot <- instring(".",gcm)
  if (length(dot)>=3) {
    run<- substr(gcm,dot[4]+1,dot[4]+4)
    gcm.nm <- substr(gcm,dot[2]+1,dot[3]-1)
  } else {
      dash <- instring("_",gcm)
      gcm.nm <- substr(gcm,1,dash[1]-1)
      run<- substr(gcm,dash[1]+1,dash[2]-1)
  }

  slsh <- instring("/",obs$location)
  subdir<-paste(predictand,strip(obs$location),obs$station,ele,sep="")
  fname <- paste(op.path,"/",subdir,"/ds_one_",ipcc.ver,
                predictand,strip(obs$location),
                obs$station,ele,".",gcm.nm,scen,".",run,".",sep="")

  if (!silent) print(fname)
```

```
  if (!file.exists(paste(fname,"txt",sep="")) &
      !file.exists(paste(fname,"Rdata",sep="")) | (force)) {
    GCM <- retrieve.nc(paste(dir,gcm,sep=""),x.rng=x.rng,
                      y.rng=y.rng,v.nam=v.nam,silent=TRUE)
    class(GCM) <- c("field","monthly.field.object")
    attr(GCM$tim,"unit") <- "month"; GCM$dd[] <- 15

    ds.station <- objDS(era40.t2m,GCM,obs,plot=FALSE,
                       qualitycontrol=qc,silent=silent)
    x <- ds.station$station
    x$grade.pattern <- min(
c(ds.station$Jan$grade.pattern,ds.station$Feb$grade.pattern,
  ds.station$Mar$grade.pattern,ds.station$Apr$grade.pattern,
  ds.station$May$grade.pattern,ds.station$Jun$grade.pattern,
  ds.station$Jul$grade.pattern,ds.station$Aug$grade.pattern,
  ds.station$Sep$grade.pattern,ds.station$Oct$grade.pattern,
  ds.station$Nov$grade.pattern,ds.station$Dec$grade.pattern))
    x$grade.trend <-   min(c(
  ds.station$Jan$grade.trend,ds.station$Feb$grade.trend,
  ds.station$Mar$grade.trend,ds.station$Apr$grade.trend,
  ds.station$May$grade.trend,ds.station$Jun$grade.trend,
  ds.station$Jul$grade.trend,ds.station$Aug$grade.trend,
  ds.station$Sep$grade.trend,ds.station$Oct$grade.trend,
  ds.station$Nov$grade.trend,ds.station$Dec$grade.trend))
    x.ts <- plotStation(x,what="n",add=TRUE,col="steelblue",
                       type="l",lwd=2,lty=1,l.anom=FALSE,
                       mon=cmons,trend=TRUE,std.lev=FALSE)
    if (!test) x$val <- x$val - x.ts$trend[1] +
              obs.ts$trend[length(obs.ts$trend)]
    print(paste("Saving in",fname))

    if (!file.exists( paste(op.path,"/",subdir,sep="") )) {
        if (!silent) print(paste("Creating (3)",op.path,"/",
                                 subdir,sep=""))
        dir.create( paste(op.path,"/",subdir,sep="") )
    } else if (!silent) print(paste(op.path,"/",subdir,
                           " exsists...",sep=""))
    save(file=paste(fname,"Rdata",sep=""),x,ds.station)
    ds.scen <- data.frame(Year=x$yy,
                   Jan=round(x$val[,1],2),Feb=round(x$val[,2],2),
                   Mar=round(x$val[,3],2),Apr=round(x$val[,4],2),
                   May=round(x$val[,5],2),Jun=round(x$val[,6],2),
                   Jul=round(x$val[,7],2),Aug=round(x$val[,8],2),
                   Sep=round(x$val[,9],2),Oct=round(x$val[,10],2),
                   Nov=round(x$val[,11],2),Dec=round(x$val[,12],2))
```

```
    write.table(ds.scen,file=paste(fname,"txt",sep=""),
                row.names = FALSE,quote = FALSE, sep="\t ")
  } else {
    print(paste(fname,"exists, skipping ESD for this GCM."))
         load(paste(fname,"Rdata",sep=""))
  }
      plotStation(x,what="t",add=TRUE,col="steelblue",type="l",lwd=2,
                lty=1,l.anom=FALSE,mon=cmons,trend=TRUE,std.
                lev=FALSE)

}

print("Add observations...")
plotStation(obs,what="t",add=TRUE,col="grey20",type="p",pch=19,
           l.anom=FALSE,mon=cmons,trend=TRUE,std.lev=FALSE)
print("Finished plotting")

if (options()$device=="X11") {
  dev.copy2eps(file=paste("ds_one_",ipcc.ver,obs$location,obs$ele,
                          ".eps",sep=""))
  dev2bitmap(file=paste("ds_one_",ipcc.ver,obs$location,obs$ele,
                        ".png",sep=""))
}
if (options()$device=="png") dev.off()

 if (post) {
   a <- Sys.info()
   script <- rep("",12)
   script[1] <- "#!/usr/bin/ksh"
   script[2] <- paste("echo putftp ",fname.png)
   script[3] <- "echo FTP: put data file to lightning.."
   script[4] <- "cat<<eof | ftp -in"
   script[5] <- "open lightning"
   script[6] <- paste("user",a[7],passwd)
   script[7] <- "cd pub_ftp"
   script[8] <- "pwd"
   script[9] <- paste("mput ",fname.png)
   script[10] <- "ls"
   script[11] <- "quit"
   script[12] <- "eof"
   writeLines(file="ftpscript.sh",script)
   system("chmod a+x  ftpscript.sh")
   system("./ftpscript.sh")
 }
 if (test) invisible(x)
print("HERE")
}
```

REFERENCES

Abaurrea, J. and Asín, J., Forecasting local daily precipitation patterns in a climate change scenario, *Clim. Res.*, **28** (2005) 183–197.

Alexander, L., Zhange, X., Peterson, T., Caesar, J., Gleason, B. E., Klein Tank, A. M. G., Haylock, M. R., Collins, M. and Trewin, B., Global observed changes in daily climate extremes on temperature and precipitation, *J. Geophys. Res.*, in press.

Anderson, D. L. T. and Carrington, D., Simulation of tropical variability as a test of climate models, in *Global Change. Environment and Quality of Life,* eds. Speranza, A., Fantechi, R. and Tibaldi, S., Vol. EUR 15158 EN.

Anderson, T. W., *An Introduction to Multivariate Statistical Analysis*, 1st edn. (John Wiley & Sons, New York, 1958), 374 pp.

Arnell, N. and Osborn, T., Interfacing climate and impact models in integrated assessment modelling, *Technical Report 52* (Tyndall Centre for Climate Change Research, 2006).

Bailey, N. T. J., *The Elements of Stochastic Processes with Applications to the Natural Sciences* (Wiley, 1964), 249 pp.

Baker, D. G., Synoptic-scale and mesoscale contributions to objective operational maximum-minimum temperature forecast errors, *Monthly Weather Rev.*, **110** (1982) 163–169.

Barnett, T. P., Comparison of near-surface air temperature variability in 11 coupled global climate models, *J. Clim.*, **12** (1999) 511–518.

Behera, S., Rao, S. A., Saji, H. N. and Yamagata, T., Comments on "A cautinoary note on the interpretation of EOFs," *J. Clim.*, **16** (2003) 1087–1093.

Benestad, R. E., CCA applied to statistical downscaling for prediction of monthly mean land surface temperatures: Model documentation, *Klima 28/98* (DNMI, Oslo, Norway, 1998a).

Benestad, R. E., SVD applied to statistical downscaling for prediction of monthly mean land surface temperatures: Model documentation, *Klima 30/98* (DNMI, Oslo, Norway, 1998b).

Benestad, R. E., Evaluation of seasonal forecast potential for Norwegian land temperatures and precipitation using CCA, *Klima 23/99* (DNMI, Oslo, Norway, 1999a).

Benestad, R. E., MVR applied to statistical downscaling for prediction of monthly mean land surface temperatures: Model documentation, *Klima 2/99* (DNMI, Oslo, Norway, 1999b).

Benestad, R. E., Pilot studies of enhanced greenhouse gas scenarios for Norwegian temperature and precipitation from empirical downscaling, *Klima 16/99* (DNMI, Oslo, Norway, 1999c).

Benestad, R. E., S-mode and T-mode EOFs from a GCM modeller's perspective: Notes on the linear algebra, *Klima 24/99* (DNMI, Oslo, Norway, 1999d).

Benestad, R. E., Analysis of gridded sea level pressure and 2-meter temperature for 1873–1998 based on UEA and NCEP re-analysis II, *KLIMA 03/00* (DNMI, Oslo, Norway, 2000).

Benestad, R. E., The cause of warming over Norway in the ECHAM4/OPYC3 *GHG* integration, *Int. J. Climatol.* **21** (2001a) 371–387.

Benestad, R. E., A comparison between two empirical downscaling strategies, *Int. J. Climatol.*, **21** (2001b) 1645–1668, doi:10.1002/joc.703.

Benestad, R. E., Empirically downscaled multi-model ensemble temperature and precipitation scenarios for Norway, *J. Clim.*, **15** (2002a) 3008–3027.

Benestad, R. E., Empirically downscaled temperature scenarios for northern Europe based on a multi-model ensemble, *Clim Res.*, **21** (2002b) 105–125.

Benestad, R. E., *Solar Activity and Earth's Climate* (Springer, Berlin, 2002c), 287 pp.

Benestad, R. E. *clim.pact-V.1.0, KLIMA 04/03* (The Norwegian Meteorological Institute, Oslo, Norway, 2003a), www.met.no.

Benestad, R. E., *clim.pact-V.1.0, KLIMA 04/03* (The Norwegian Meteorological Institute, Oslo, Norway, 2003b), www.met.no.

Benestad, R. E., Downscaling analysis for daily and monthly values using *clim.pact-V.0.9, KLIMA 01/03*, met.no (The Norwegion Meteorological Institute, Oslo, Norway, 2003c), www.met.no.

Benestad, R. E., How often can we expect a record-event?, *Clim. Res.*, **25** (2003d) 3–13.

Benestad, R. E., What can present climate models tell us about climate change?, *Clim. Change*, **59** (2003e) 311–332.

Benestad, R. E., Are temperature trends affected by economic activity? Comment on McKitrick and Michaels, *Clim. Res.*, **27** (2004a) 171–173.

Benestad, R. E., Empirical-statistical downscaling in climate modeling, Eos *Trans. Am. Geophys. Union*, **85**(42) (2004b) 417.

Benestad, R. E., Empirically downscaled SRES-based climate scenarios for Norway, in *Climate 08* (The Norwegian Meteorological Institute, Oslo, Norway, 2004c), www.met.no.

Benestad, R. E., Record-values, non-stationarity tests and extreme value distributions, *Global Planetary Change*, **44** (2004d) 11–26, doi:10.1016/j.gloplacha.2004.06.002.

Benestad, R. E., Tentative probabilistic temperature scenarios for northern Europe, *Tellus* **56A** (2004e) 89–101.

Benestad, R. E., Climate change scenarios for northern Europe from multi-model IPCC AR4 climate simulations, *Geophys. Res. Lett.*, **32** (2005) L17704, doi:10.1029/2005GL023401.

Benestad, R. E., Novel methods for inferring future changes in extreme rainfall over northern Europe, *Clim. Res.*, **34** (2007) 195–210, doi:10.3354/cr00693.

Benestad, R. E. and Chen, D., The use of a calculus-based cyclone identification method for generating storm statistics, *Tellus A*, **58A** (2006) 473–486, doi:10.1111/j.1600-0870.2006.00191.x.

Benestad, R. E. and Melsom, A., Is there a link between the unusually wet autumns in southeastern Norway and SST anomalies? *Clim. Res.*, **23** (2002) 67–79.

Benestad, R. E., Hanssen-Bauer, I., Førland, E. J., Tveito, O. E. and Iden, K., Evaluation of monthly mean data fields from the ECHAM4/OPYC3 control integration, *Klima 14/99* (DNMI, Oslo, Norway, 1999).

Benestad, R. E., Hanssen-Bauer, I. and Førland, E. J., Empirically downscaled temperature scenarios for Svalbard, *Atmos. Sci. Lett.*, **3**(2–4) (2002) 71–93, doi:10.1006/asle.2002.0051.

Benestad, R. E., Achberger, C. and Fernandez, E., Empirical-statistical downscaling of distribution functions for daily precipitation, Climate 12/2005 (The Norwegian Meteorological Institute), www.met.no.

Benestad, R. E., Hanssen-Bauer, I. and Førland, E. J., An evaluation of statistical models for downscaling precipitation and their ability to capture long-term trends, *Int. J. Climatol.*, **27** (2007) 649–665, doi:10.1002/joc.1421.

Bengtsson, L., The climate response to the changing greenhouse gas concentration in the atmosphere, in *Decadal Variability*, eds. Anderson, D. L. T. and Willebrand, J., NATO ASI series, Vol. 44 (1996).

Bergant, K. and Kajfež-Bogataj, L., N-PLS regression as empirical downscaling tool in climate change studies, *Theo. Appl. Climatol.*, **81** (2005) 11–23.

Bergant, K., Kajfež-Bogataj, L. and Črepinšek, Z., The use of EOF analysis for preparing the phenological and climatological data for statistical downscaling — Case study: The beginning of flowering of the Dandelion (*Taraxacum officinale*) in Slovenia, in *Developments in Statistics*, (eds.) Mrvar, A. and Ferligoj, A., pp. 163–174.

Bretherton, C. S, Smith, C. and Wallace, J. M., An intercomparison of methods for finding coupled patterns in climate data, *J. Clim.* **5** (1992) 541–560.

Brink, K. H. and R. O. Muench, Circulation in the point conception-Santa Barbara channel region, *J. Geophys. Res.*, **91** (1986) 877–895.

Bubuioc, A., Giorgi, F., Bi, X. and Ionita, M., Comparison of regional climate model and statistical downscaling simulations of different winter precipitation change scenarios over Romania, *Theo. Appl. Climatol.*, **86** (2006) 101–123.

Busuioc, A., Chen, D. and Hellström, C., Performance of statistical downscaling models in GCM validation and regional climate change estimates: Application for Swedish precipitation, *Int. J. Climatol.*, **21** (2001) 557–578.

Busuoic, A., von Storch, H. and Schnur, R. Verification of GCM — generated regional seasonal precipitation for current climate and of statistical downscaling estimates under changing climate conditions, *J. Clim.*, **12** (1999) 258–272.

Chen, D., A monthly circulation climatology for Sweden and its application to a winter temperature case study, *Int. J. Climatol.*, **20** (2000) 1067–1076.

Chen, D. and Chen, Y., Association between winter temperature in China and upper air circulation over East Asia revealed by *Canonical Correlation Analysis, Global and Planetary Change* **37** (2003) 315–325.

Chen, D. and Hellström, C., The influence of the North Atlantic oscillation on the regional temperature variability in Sweden: Spatial and temporal variations, *Tellus*, **51A** (1999) 505–516.

Chen, D., Achberger, C., Räisänen, J. and Hellström, C., Using statistical downscaling to quantify the GCM-related uncertainty in regional climate change scenarios: A case study of Swedish precipitation, *Adv. Atmos. Sci.*, **23** (2005) 54–60.

Chen, D., Gong, L., Xu, C. Y. Halldin, S., A high-resolution, gridded dataset for monthly temperature normals (1971–2000) in Sweden, *Geografiska Annaler.* **Series 89A**(4) (2007) 249–261.

Chen, D., Walter, A., Moberg, A., Jones, P. D., Jacobeit, J., Lister, D., Trend atlas of the EMULATE indices, Research Report C73 (Earth Sciences Centre, University of Gothenburg, Gothenburg, Sweden) (2006) 797 p, http://www.gvc2.gu.se/ngeo/rcg/pdf/EmulateTrdAtlas2006_Final20070111.pdf.

Christensen, J. H. and Christensen, O. B., Severe summertime flooding in Europe, *Nature*, **421** (2002) 805.

Christensen, J. H., Risnen, J., Iversen, T., Bjørge, D., Christensen, O. B. and Rummukainen, A synthesis of regional climate change simulations — A Scandinavian perspective, *Geophys. Res. Lett.*, **28**(6) (2001) 1003.

Christensen, J. H., Hewitson, B., Busuioc, A., Chen, A., Gao, X., Held, I., Jones, R., Kolli, R. K., Kwon, W.-T., Laprise, R., na Rueda, V., Maga Mearns, L., Menéndez, C. G., Räisänen, J., Rinke, A., Sarr, A. and Whetton, P., Regional climate projections, in *Climate Change: The Physical Science Basis* (Cambridge University Press, United Kingdom and New York, NY, USA).

Christensen, O. B., Christensen, J. H., Machenhauer, B. and Botzet, M., Very high-resolution climate simulations over Scandinavia — Present climate, *J. Clim.*, **11** (1998) 3204–3229.

Clark, M. P., Gangopadhyay, S., Brandon, D., Werner, K., Hay, L., Rajagopalan, B. and Yates, D., A resampling procedure for generating conditioned daily weather sequences, *Water Resour. Res.*, **40**(W04304) (2004), doi:10.1029/2003WR002747.

Corte-Real, J., Qian, B. and Xu, H., Regional climate change in Portugal: Precipitation variability associated with large-scale atmospheric circulation, *Int. J. Climatol.*, **18** (1998) 619–635.

Crane, R. G. and Hewitson, B. C., Doubled CO_2 precipitation changes for the Susquehanna basin: Downscaling from the genesis general circulation model, *Int. J. Climatol.*, **18** (1998) 65–76.

Das, L. and Lohar, D., Construction of climate change scenarios for a tropical monsoon region, *Clim. Res.*, **30** (2005) 39–52.

DeGaetano, A. T. and Allen, R. J., Trends in twentieth-century extremes across the United States, *J. Clim.*, **15** (2002) 3188–3205.

Dehn, M., Application of an analog downscaling technique to the assessment of future landslide activity — A case study in the Italian Alps, *Clim. Res.*, **13** (1999) 103–113.

Easterling, D. R., Development of regional climate scenarios using a downscaling approach, *Clim. Change*, **41** (1999) 615–634.

Ellner, S. P., Review of R, Version 1.1.1. *Bull. Ecol. Soc. Am.* **82** (2001) 127–128.

Engen-Skaugen, T., Refinement of dynamically downscaled precipitation and temperature scenarios, *Climate 15/2004. met.no*, www.met.no.

Feddersen, H. and Andersen, U., A method for statistical downscaling of seasonal ensemble predctions, *Tellus*, **57A** (2005) 398–408.

Fernandez, J. and Saenz, J., Improved field reconstruction with the analog method: Searching the CCA space, *Clim. Res.*, **24** (2003) 199–213.

Ferro, C. A. T., Hannachi, A. and Stephenson, D. B., Simple nonparamteric techniques for exploring changing probability distributions of weather, *J. Clim.*, **18** (2005) 4344–4354.

Feuerstein, B., Dotzek, N. and Grieser, J., Assessing a Tornado climatology from global Tornado intensity distributions, *J. Clim.*, **18** (2005) 585–596.

Flury, B., Common principal components and related multivariate models, in *Wiley Series in Probability and Mathematical Statistics* (Wiley, 1988).

Fowler, H. J., Blenkinsop, S. and Tebaldi, C., Linking climate change modelling to impacts studies: Recent advances in downscaling techniques for hydrological modelling, *Int. J. Climatol.*, **27** (2007) 1547–1578.

Frich, P., Alexandersson, H., Ashcroft, J., Dahlström, B., Demarée, G. R., Drebs, A., van Engelen A. F. V, Førland, E. J., Hanssen-Bauer, I., Heino, R., Jónsson, T., Jonasson, K., Nordli, P. Ø., Schmidth, T., Steffensen, P., Tuomenvirta, H. and Tveito, O. E., North Atlantic Climatological Dataset (NACD Version 1) — Final Report, *Scientific Report 1* (DMI, Copenhagen, Denmark, 1996).

Frich, P., Alexander, L. V., Della-Marta, P., Gleason, B., Haylock, M., Klein Tank, A. M. G. and Peterson, T., Observed coherent changes in climatic extremes during the second half of the twentieth century, *Clim. Res.*, **19** (2002) 193–212.

Gabriel, R. and Neumann, J., A Markov chain model for daily rainfall occurrence in Tel Aviv Israel, *Q.J.R. Met. Soc.*, **88** (1962) 90–95.

Gentleman, R. and Ihaka, R., Lexical scope and statistical computing, *J. Comput. Graph. Stat.*, **9** (2000) 491–508.

Gill, A. E., *Atmosphere-Ocean Dynamics* (Elsevier, 1982).

Gleick, J., *Chaos* (Penguin (Non-classics) 1988).

Goodess, C. M. and Palutikof, J. P., Development of daily rainfall scenarios for southeast Spain using a circulation-type approach to downscaling, *Int. J. Climatol.*, **10** (1998) 1051–1083.

Graham, Ph., Chen, D., Christensen, O. B., Kjellström, E., Krysanova, V., Meier, M., Radziejewski, M., Räisänen, J., Rockel, B., Ruosteenoja, K., Chapter 3 of *The BALTEX Assessment of Climate Change for the Battic Sea Basin*: Projections of future climate change (Springer, 2008).

Grotch, S. and MacCracken, M., The use of general circulation models to predict regional climate change, *J. Clim.*, **4** (1991) 286–303.

Hanssen-Bauer, I., Downscaling of temperature and precipitation in Norway based upon multiple regression of the principal components of the SLP field, *KLIMA 21/99* (DNMI, 1999).

Hanssen-Bauer, I. and Førland, E., Temperature and precipitation variations in Norway 1900–1994 and their links to atmospheric circulation, *Int. J. Climatol.*, **20** (2000) 1693–1708.

Hanssen-Bauer, I., Førland, E. J., Haugen, J. E. and Tveito, O. E., Temperature and precipitation scenarios for Norway: Comparison of results from dynamical and empirical downscaling, *Clim. Res.*, **25** (2003) 15–27.

Hanssen-Bauer, I., Achberger, C., Benestad, R. E., Chen, D. and Førland, E. J., Statistical downscaling of climate scenarios over Scandinavia: A review, *Clim. Res.*, **29** (2005) 255–268.

Hayhoe, K., Cayan, D., Field, C. B., Frumhoff, P. C., Maurer, E. P., Miller, N. L., Moser, S. C., Schneider, S. H., Cahill, K. N., Cleland, E. E., Dale, L., Drapek, R., Hanemann, R. M., Kalkstein, L. S., Lenihan, J., Lunch, C. K., Neilson, R. P., Sheridan, S. C. and Verville, J. H., Emission pathways, climate change, and impacts on California. PNAS, Vol. 101, pp. 12422–12427.

Hellström, C., Chen, D., Achberger, C. and Räisanen, J., Comparison of climate change scenarios for Sweden based on statistical and dynamical downscaling of monthly precipitation, *Clim Res.*, **19** (2001) 45–55.

Hewitson, B. C. and Crane, R. G., Self-organizing maps: Applications to synoptic climatology, *Clim. Res.*, **22** (2002) 13–26.

Heyen, H., Zorita, E. and von Storch, H., Statistical downscaling of monthly mean North Atlantic air-pressure to sea level anomalies in the Baltic Sea, *Tellus*, **48A** (1996) 312–323.

Höfer, T., Przyrembel, H. and Verleger, S., New evidence for the Theory of the Stork, *Paediat. Perinatal Epidemiolo.*, **18** (2004) 88–92, doi:10.1111/j.1365-3016.2003.00534.x.

Horton, E. B., Folland, C. K. and Parker, D. E., The changing incidence of extremes in worldwide and central England temperature to the end of the twentieth century, *Clim. Change* **50** (2001) 267–295.

Houghton, J. T., Ding, Y., Griggs, D. J., Noguer, M., van der Linden, P. J., Dai, X., Maskell, K. and Johnson, C. A. Climate change 2001: The scientific basis, Contribution of Working Group I to the Third Assessment Report of IPCC, *Int. Panel Climate Change* (Available from www.ipcc.ch).

Hulme, M. and Jenkins, G. J., Climate change scenarios for the United Kingdom, *Scientific Report 1/98* (UK Met Office).

Huntingford, C., Jones, R. G., Prudhomme, C., Lamb, R., Gash, J. H. C. and Jones, D. A., Regional climate-model predictions of extreme rainfall for a changing climate, *Quart. J. R. Met. Soc.*, **129** (2003) 1607–1621, doi:10.1256/gj.0297.

Huth, R., Statistical downscaling of daily temperature in central Europe, *J. Clim.*, **15** (2002) 1731–1742.

Huth, R., Sensitivity of local daily temperature change estimates to the selection of downscaling models and predictors, *J. Clim.*, **17** (2004) 640–652.

Huth, R. and Kyselý, J., Constructing site-specific climate change scenarios on a monthly scale, *Theor. Appl. Climatol.*, **66** (2000) 13–27.

Imbert, A., The analog method applied to downscaling of climate scenarios, *KLIMA 08/03* (The Norwegian Meteorological Institute, Oslo, Norway, 2003), www.met.no.

Imbert, A. and Benestad, R. E., An improvement of analog model strategy for more reliable local climate change scenarios, *Theo. Appl. Climatol.*, **82** (2005) 245–255, doi:10.1007/s00704-005-0133-4.

IPCC, (Beijing 11–13 June). Intergovernmental Panel on Climate Change Workshop on Changes in Extreme Weather and Climate Events. Tech. Rept. WMO/UNEP, URL: http://www.ipcc.ch/.

Johnson, E. S. and McPhaden, M. J., Structure of intraseasonal Kelvin waves in the equatorial Pacific Ocean, *J. Phys. Oceanogra.*, **23** (1993) 608–625.

Joliffe, I. T., A cautinary note on artificial examples of EOFs, *J. Clim.*, **16** (2003) 1084–1086.

Kaas, E. and Frich, P., Diurnal temperature range and cloud cover in the Nordic countries: Observed trends and estimates for the future, *Atmos. Res.*, **37** (1995) 211–228.

Kaas, E., Christensen, O. B. and Christensen, J. H., Dynamical versus empirical downscaling, Personal communication, 1998.

Kaiser, H. F., The varimax criterion for analytic rotation in factor analysis, *J. Psychomet.*, **23** (1958) 187–200, doi:10.1007/BF02289233.

Kao, C. Y. J., Langley, D. L., Reisner, J. M. and Smith, W. S., Development of the first nonhydrostatic nested-grid grid-point global atmospheric modeling system on parallel machines, *Tech. Rept. LA-UR-98-2231* (Los Alamos National Lab., NM).

Katz, R. W. and Brown, B. G., Extreme events in a changing climate: Variability is more important than averages, *Clim. Change*, **21** (1992) 289–302.

Kidson, J. W. and Thompson, C. S., A comparison of statistical and model-based downscaling techniques for estimating local climate variations, *J. Clim.*, **11** (1998) 735–753.

Kilsby, C. G., Cowpertwait, P. P. P., O'Connel, P. E. and Jones, P. D., Predicting rainfall statistics in England and Wales using atmospheric circulation variables, *Int. J. Climatol.*, **18** (1998) 523–539.

Kim, J.-W., Chang, J.-T., Baker, N. L., Wilks, D. S. and Gates, W. L., The statistical problem of climate inversion: Determination of the relationship between local and large-scale climate, *Monthly Weather Rev.*, **112** (1984) 2069–2077.

Klein, W. H., Winter precipitation as related to the 700-millibar circulation, *Bull. Amer. Meteor. Soc.*, **9** (1948) 439–453.

Klein, W. H. and Bloom, H. J., Specification of monthly precipitation over the United States from the surrounding 700 mb height field, *Mon. Wea. Rev.*, **115** (1987) 2118–2132.

Klein, W. H., Lewis, B. M. and Enger, I., Objective prediction of five-day mean temperatures during winter, *J. Meteorol.*, **16** (1959) 672–681.

Kundu, P. K. and Allen, J. S., Some three-dimensional characteristics of the low-frequency current fluctuations near the Oregon coast, *J. Phys. Oceanogra.*, **6** (1976) 181–199.

Kyselý, J., Comparison of extremes in GCM-simulated, downscaled and observed central — European temperature series, *Clim. Res.*, **20** (2002a) 211–222.

Kyselý, J., Temporal fluctuations in heat waves at Prague-klementinum, the Czech Republic, from 1901–1997, and their relationships to atmospheric circulation, *IJC*, **22** (2002b) 33–50.

Lapp, S., Byrne, J., Kienzle, S. and Townshend, I., Linking global circulation model synoptics and precipitation for western North America, *Int. J. Climatol.*, **22** (2002) 1807–1817.

Leroy, S. S., Detecting climate signals: Some Bayesian aspects, *J. Clim.*, **11** (1998) 640–651.

Liao, Y., Zhang, Q. and Chen, D., Stochastic modeling of daily precipitation in China, *J. Geograph. Sci.*, **14**(4) (2004) 417–426.

Linderson, M.-L., Achberger, C. and Chen, D., Statistical downscaling and scenario construction of precipitation in Scania, southern Sweden, *Nordic Hydrol.*, **35** (2004) 261–278.

Lorenz, E., Deterministic nonperiodic flow, *J. Atmos. Sci.*, **20** (1963) 130–141.

Lorenz, E., The nature and theory of the general circulation of the atmosphere, *Publication 218* (WMO, 1967).

Lorenz, E. N., Empirical orthogonal functions and statistical weather prediction, *Sci. Rep. 1* (Department of Meteorology, MIT, USA, 1956).

Machenhauer, B., Windelband, M., Botzet, M., Christensen, J. H., Déqué, M., Jones, R. G., Ruti, P. M. and Visconti, G., Validation and analysis of regional present-day climate and climate change simulations over Europe, *Tech. Rept. 275* (Max Planck-Institute für Meteorologie, 1998).

Mahadevan, A. and Archer, D., Modeling a limited region of the ocean, *J. Comp. Phys.*, **145** (1998) 555–574.

Mardia, K., Kent, J. and Bibby, J., *Multivariate Analysis* (Academic Press, Inc., 1979).

Matheron, G., Principles of geostatistics, *Economic Geol.*, **58** (1963) 1246–1266.

Moberg, A., Jones, P. D., Lister, D., Walther, A., Brunet, M., Jacobeit, J., Saladie, O., Aguilar, J., Sigroand, E., Della-Marta, P., Luterbacher, J., Yiou, P., Alexander, L. V., Chen, D., Klein Tank, A. M. G., Alexandersson, H., Almarza, C., Auer, I., Barriendos, M., Begert, M., Bergström, H., Böhm, R., Butler, J., Caesar, J., Drebs, A., Founda, D., Gerstengarbe, F.-W., Giusi, M., Jónsson, T., Maugeri, M., Österle, H., Pandzic, K., Petrakis, M., Srnec, L., Tolasz, R., Tuomenvirta, H., Werner, P. C., Linderholm, H., Philipp, A., Wanner, H. and Xoplaki, E., Indices for daily temperature and precipitation extremes in Europe analysed for the period 1901–2000, *J. Geophys. Res.*, **111**(D22106) (2006), doi:10.1029/2006JD007103.

Murphy, J., An evaluation of statistical and dynamical techniques for downscaling local climate, *J. Clim.*, **12** (1999) 2256–2284.

Murphy, J., Predictions of climate change over Europe using statistical and dynamical downscaling techniques, *Int. J. Clim.*, **20** (2000) 489–501.

North, G. R., Bell, T. L. and Cahalan, R. F., Sampling errors in the estimation of empirical orthogonal functions, *Monthly Weather Rev.*, **110** (1982) 699–706.

Osborne, T. J. and Jones, P. D., Air flow influences on local climate: Observed United Kingdom climate variations, *Atmos. Sci. Lett.*, **1**(1) (2004) 62–74.

Osborne, T. J., Conway, D., Hulme, M., Gregory, J. M. and Jones, P. D., Air flow influences on local climate: Observed and simulated mean relationships for the United Kingdom, *Clim Res.*, **13** (1999) 173–191.

Oshima, N., Kato, H. and Kadokura, S., An application of statistical downscaling to estimate surface air temperature in Japan, *J. Geophys. Res.*, **107**(D10) (2002) ACL–14.

Palmer, T. N., Predictability of the atmosphere and oceans: From days to decades, in *Decadal Variability*, eds. Anderson, D. L. T. and Willebrand, J., NATO ASI Series, Vol. 44, 1996.

Palmer, T. N. and Räisänen, J., Quantifying the risk of extreme seasonal precipitation events in a changing climate, *Nature*, **415** (2002) 512–514.

Peixoto, J. P. and Oort, A. H., *Physics of Climate* (American Institute Physics, 1992), 520 p.

Penlap, E. K., Matulla, M., von Storch, H. and Kamga, F. M., Downscaling of GCM scenarios to assess precipitation changes in the little rainy season (March–June) in Cameroon, *Clim. Res.*, **26** (2004) 85–96.

Philander, S. G., *El Niño, La Niña, and the Southern Oscillation* (Academic Press, 1989), 287 pp.

Preisendorfer, R. W., *Principal Component Analysis in Meteorology and Oceanology* (Elsevier Science, 1988).

Press, W. H., Flannery, B. P., Teukolsky, S. A. and Vetterling, W. T., *Numerical Recipes in Pascal* (Cambridge University Press, 1989).

Prudhomme, C. and Reed, D., Mapping extreme rainfall in a mountainous region using geostatistical techniques: A case study in Scotland, *Int. J. Climatol.*, **19** (1999) 1337–1356.

Pryor, S. C., School, J. T. and Barthelmie, R. J., Climate change impacts on wind speeds and wind energy density in northern Europe: Empirical downscaling of multiple AOGCMs, *Clim. Res.*, **29** (2005) 183–198.

Pryor, S. C., School, J. T. and Barthelmie, R. J., Winds of change? Projections of near-surface winds under climate change scenarios, *Geophys. Res. Lett.*, **33** (2006), doi:10.1029/2006GL026000.

Pryor, S. C., School, J. T. and Barthelmie, R. J., Empirical downscaling of wind speed probability distributions, *J. Geophys. Res.*, submitted for publication.

Racsko, P., Szeidl, L. and Semenov, M. A., A serial approach to local stochastic weather models, *Ecolog. Modell.*, **57** (1991) 27–41.

Räisänen, J., Rummukainen, M., Ullerstig, A., Bringfelt, B., Hansson, U. and Willén, U., The first Rossby Centre Regional Climate Scenario — Dynamical downscaling of CO_2-induced climate change in the HadCM2 GCM, SWECLIM 85. SMHI.

Reason, C. J. C., Landman, W. and Tennantc, W., Seasonal to decadal prediction of southern African climate and its links with variability of the Atlantic Ocean, *Bull. Amer. Meteor. Soc.* **87** (2006) 941–955.

Richardson, C., Stochastic simulation of daily precipitation, temperature, and solar radiation, *Water Resou. Res.*, **17** (1981) 182–190.

Robinson, P. J. and Finkelstein, P. L., The development of impact-oriented scenarios, *Bull. Am. Met. Soc.*, **4** (1991) 481–490.

Rummukainen, M., Methods for statistical downscaling of GCM simulations, SWECLIM 80, SMHI.

Rummukainen, M., Räsänen, J. and Graham, P., Regional climate simulations for the Nordic region — First results from SWECLIM, SWECLIM Nov. 1998, SMHI.

Salathé, E. P., Downscaling simulations of future global climate with application to hydrologic modelling, *Int. J. Climatol.*, **25** (2005) 419–436.

Sarachik, E. S., Winton, M. and Yin, F. L., Mechanisms for decadal-to-centennial climate variability, in *Decadal Variability*, eds. Anderson, D. L. T. and Willebrand, J., NATO ASI series, Vol. 44, 1996.

Satoh, M., *Atmospheric Circulation Dynamics and General Circulation Models* (Springer, 2004).

Schoof, J. T. and Pryor, S. C., Downscaling temperature and precipiation: A comparison of regression-based methods and artificial neural networks, *Int. J. Climatol.*, **21** (2001) 773–790.

Schubert, S., Downscaling local extreme temperature change in south-eastern Australia from the CSIRO MARK2 GCM, *Int. J. Climatol.*, **18** (1998) 1419–1438.

Semenov, M. and Barrow, E. M., Use of a stochastic weather generator in the development of climate change scenarios, *Clim. Change*, **35** (1997) 397–414.

Semenov, M. A., Development of high-resolution UKCIP02-based climate change scenarios in the UK, *Agric. Forest Meteorol.*, **144** (2007) 127–138.

Semenov, M. A. and Brooks R. J., Spatial interpolation of the LARS-WG stochastic weather generator in Great Britain, *Clim. Res.*, **11** (1999) 137–148.

Sengupta, S. and Boyle, J. S., Using common principal components in comparing GCM simulations, *J. Clim.*, **11** (1998) 816–830.

Sengupta, S. K. and Boyle, J. S., Statistical intercomparison of global climate models: A common principal component approach, *Tech. Rept. 13. PCMDI* (Lawrence Livermore National Laboratory, California, USA, 1993), http://www-pcmdi.llnl.gov/pcmdi/pubs/pdf/13.pdf.

Skaugen, T., Astrup, M., Roald, L. A. and Skaugen, T. E., Scenarios of extreme precipitation of duration 1 and 5 days for Norway caused by climate change, *Tech. Rept.* (Norges vassdrags- og energidirektorat, NVE, Oslo, Norway, 2002a).

Skaugen, T. E., Hanssen-Bauer, I. and Førland, E. J., Adjustment of dynamically downscaled temperature and precipitation data in Norway, *KLIMA 20/02* (The Norwegian Meteorological Institute, Oslo, Norway, 2002b), www.met.no.

Sneyers, R., On statistical analysis of series of observations, *Technical Note 143* (WMO, Geneva, Switzerland, 1990).

Soltani, A. and Hoogenboom, G., Minimum data requirements for parameter estimation of stochastic weather generators, *Clim. Res.*, **25** (2003) 109–119.

Srikanthan, R. and McMahon, T. A., Stochastic generation of annual, monthly and daily climate data: A review, *Hydrol. Earth Syst. Sci.*, **5**(4) (2001) 653–670.

Stephenson, D. B., Kumar, K. R., Doblas-Reyes, F. J., Royer, J. F., Chauvin, F. and Pezzulli, S., Extreme daily rainfall events and their impact on estimating the predictability of the Indian Monsoon, *Note de Centre* 63 (Meteo France, Centre National De Recherches, Meteorologiques, 1998).

Strang, G., *Linear Algebra and its Application* (Harcourt College Pub; 4 Tch edition, 1988).

Sumner, G. N., Romero, R., Homar, V., Ramis, C., Alonso, S. and Zorita, E., An estimate of the effects of climate change on the rainfall of Mediterranean Spain by the late twenty first century, *Clim. Dyn.*, **20** (2003) 789–805, doi:10.1007/s00382-003-0307-7.

Timbal, B., Dufour, A. and McAvaney, B., An estimate of future climate change for western France using a statistical downscaling technique, *Clim. Dyn.*, **20** (2003) 807–823, doi:10.1007/s00382-002-0298-9.

Treut, H. Le., Cloud representation in large-scale models: Can we adequately constrain them through observed data? in *Global Change. Environment and Quality of Life,*, eds. Speranza, A., Fantechi, R. and Tibaldi, S., vol. EUR 15158 EN.

Tubiello, F., Donatelli, M., Rozenweig, C. and Stckle, C. O., Effects of climate change and elevated CO_2 on cropping systems: Model predictions at two Italian locations, *Eur. J. Agron.*, **13** (2000) 179–189.

van den Dool, H. M., Constructed analogue prediction of the East Central tropical pacific through Spring 1996, *NOAA: Exp Long-Lead Forecast Bull.*, **4** (1995) 41–43.

van Oldenborgh, G. J., Extraordinarily mild European autumn 2006 due to global warming? *Global Change Newslett.* (2006) 18–20.

von Storch, H., On the use of "Inflation" in statistical downscaling, *J. Clim.*, **12** (1999) 3505–3506.

von Storch, H. and Zwiers, F. W., *Statistical Analysis in Climate Research.*

von Storch, H., Zorita, E. and Cubasch, U., Downscaling of global climate change estimates to regional scales: An application to Iberian rainfall in wintertime, *J. Clim.*, **6** (1993) 1161–1171.

von Storch, H., Hewitson, B. and Mearns, L., Review of empirical downscaling techniques, in *Regional Climate Development Under Global Warming*, eds. Iversen, T. and Høiskar, B. A. K., General Technical Report 4, 2000, http://regclim.met.no/rapport_4/presentation02/presentation02.htm.

Wallace, J. M. and Dickinson, R. E., Empirical orthogonal representation of time series in the frequency domain: I. Theoretical considerations, *J. Appl. Meteor.*, **11** (1972) 887–892.

Wallis, T. W. R. and Griffiths, J. F., Simulated meteorological input for agricultural models. *Agri. Forest Meteorol.*, **88** (1997) 241–258.

Wetterhall, F., Bárdossy, A., Chen, D., Halldin, S. and Xu, C.-Y., Daily precipitation downscaling techniques in different climate regions in China, *Water Resour. Res.*, **42**(W11423) (2006), doi:10.1029/2005WR004573.

Wight, J. R. and Hanson, C. L., Use of stochastically generated weather records with rangeland simulation models, *J. Range Manage.*, **44** (1991) 282–285.

Wilby, R. L., Non-stationarity in daily precipitation series: Implications for GCM down-scaling using atmospheric circulation indices, *Int. J. Climatol.*, **17** (1997) 439–454.

Wilby, R. L. and Harris, I., A framework for assessing uncertainties in climate change impacts: Low-flow scenarios for the River Thames, UK, *Water Resour. Res.*, **42** (2006), doi:10.1029/2005WR004065.

Wilby, R. L., Charles, S. P., Zortia, E., Timbal, B., Whetton, P. and Mearns, L. O. Guidelines for use of climate scenarios developed from statistical downscaling methods, Supporting Material of the Intergovernmental Panel on Climate Change (Task group on Data and Scenario Support for Impacts and Climate Analysis).

Wilby, R. L., Hassan, H. and Hanaki, K., Statistical downscaling of hydrometeorological variables using general circulation model output, *J. Hydrol.*, **205** (1998) 1–19.

Wilks, D. S. and Wilby, R. L., The weather generation game: A review of stochastic weather models, *Progress Phys. Geogra.*, **23** (1999) 329.

Wilks, D. S., Adapting stochastic weather generation algorithms for climate change studies, *Clim. Change*, **22** (1992) 67–84.

Wilks, D. S., *Statistical Methods in the Atmospheric Sciences* (Academic Press, 1995), 467 pp.

Yan, Z., Bate, S., Chandler, R. E., Isham, V. and Wheater, H., Changes in extreme wind speeds in NW Europe simulated by generalized linear models, *Theo. Appl. Climatol.*, **83** (2006) 121–137.

Yang, C., Chandler, R. E., Isham, V. S. and Wheather, H. S., Spatial-temporal rainfall simulation using generalized linear models, *Water Resour. Res.* **41** (2005), doi:10.1029/2004WR003739.

Zemansky, M. W. and Dittman, R. H., *Heat and Thermodynamics*, 6th edn. (McGraw Hill, 1981).

Zorita, E. and von Storch, H., A survey of statistical downscaling results, *Tech. Rept. 97/E/20* (GKSS, 1997).

Zorita, E. and von Storch, H., The analog method as a simple statistical downscaling technique: Comparison with more complicated methods. *J. Clim.*, **12** (1999) 2474–2489.

Zorita, E., Hughes, J. P., Lettermaier, D. P. and von Storch, H., Stochastic characterization of regional circulation patterns for climate model diagnosis and estimation of local precipitation, *J. Clim.*, **8** (1995) 1023–1042.

INDEX